FIELD GUIDE TO THE

WILD FLOWERS

OF THE

ALGARVE

FIELD GUIDE TO THE WILD FLOWERS OF THE ALGARVE

Chris Thorogood
Simon Hiscock

Kew Publishing
Royal Botanic Gardens, Kew

Kew
ROYAL BOTANIC GARDENS

First published in 2014 by
Royal Botanic Gardens, Kew
Richmond, Surrey, TW9 3AB, UK
www.kew.org

ISBN 978 1 84246 497 7

Distributed on behalf of the Royal Botanic Gardens, Kew in North America by the University of Chicago Press, 1427 East 60th Street, Chicago, IL 60637, USA.

British Library Cataloguing in Publication Data
A catalogue record for this book is available from the British Library

Editors: Sharon Whitehead, Rhian Smith
Design, typesetting and page layout: Christine Beard
Production manager: Georgina Smith

Printed in the UK by Latimer Trend

FSC www.fsc.org MIX Paper from responsible sources FSC® C013436

Contents

Foreword

The wild flowers that make up the flora of the Algarve have been living magnets for botanists both amateur and professional for a long time. 2014 will see their diversity and wonder published in the form of a stunning Field Guide that covers all aspects. My congratulations to Chris Thorogood and Simon Hiscock, School of Biological Sciences, University of Bristol.

Professor David Bellamy OBE, PhD, FLS

Acknowledgements

In preparing this Flora, we are grateful to numerous friends, colleagues and students from the School of Biological Sciences, University of Bristol, who have been involved in undergraduate field courses run in the region between 2002 and 2013. In particular, we thank Andy Bailey, David Tabah, Stephanie McInnis, Alexandra Allen and Jon Bridle. We are also especially grateful to Peter Placito and David Mabberley, whose classic work on the region's natural history: *Algarve Plants and Landscape: Passing Tradition and Ecological Change* was a source of information and inspiration. Peter, who sadly died in March 2009, helped establish the Bristol Field Course and for a number of years taught students about the region's traditional agricultural practices, a subject dear to his heart. Peter was also instrumental in establishing a field course for undergraduate students from Oxford, which David Mabberley ran from 1987 to 1997, and which first introduced SJH to the region's flora. We therefore acknowledge former colleagues from this course, especially: Quentin Cronk, Robert Scotland, Colin Hughes, Timothy Walker, Roger Hall and Jaap Vermeulen (University of Leiden). Finally, SH remembers with affection Margherita Placito (1945–2000) whose artistic and poetic appreciation of the region's landscape and flora opened many eyes to its wonders.

Species described in this book were authenticated with *Flora Europaea* and also *Flora Iberica*, which, although incomplete during the compilation of this book, was a valuable identification resource.

The Algarve

The Algarve is the southern-most province of Portugal and one of the most popular tourist destinations in Europe, receiving approximately 7 million tourists each year. Although not geographically part of the Mediterranean basin, the Algarve, with its Atlantic outlook, has a Mediterranean climate and its flora is typically Mediterranean. The region covers an area of approximately 5,500 km^2 stretching about 135 km from the windward Atlantic coast in the west (*O Barlavento*), to the more sheltered Spanish border region in the east (*O Sotavento*). Moving inland over a distance of roughly 40 km from the low-lying southern coastal zone (the *litoral*), the land rises gradually northwards through a region of rolling hills (the *barrocal*) until a somewhat discontinuous mountain range (the *serra*) forms a natural boundary with the neighbouring province of Alentejo.

The coastal city of Faro is the Algarve's provincial capital and the administrative centre to its 16 regional municipalities. Tourism is now the main driver of the Algarvian economy, and as a consequence, the region has been greatly developed in recent decades, especially along the southern coast. This development has brought relative prosperity to the region but has impacted greatly on the local flora and fauna, except in the far south west. Most tourists arrive through Faro's international airport, from where they head to holiday destinations in the east, such as Tavira, or (more usually) in the west, where the most popular resorts are found around Albufeira, Portimão and Lagos.

West of Lagos towards Sagres, the coast becomes more rocky and exposed, with fewer sheltered sandy beaches, and consequently there is less development. This remote westerly region, extending from the dramatic wind-swept cliffs of the Cape St. Vincent peninsula northwards up the Atlantic Coast and inland to the *serra* (around Monchique), contains some of the most beautiful scenery in the Algarve and offers the best opportunities to experience the region's rich ***native*** flora.

The three broad geographical regions of the Algarve: *litoral*, *barrocal* and *serra*.

CLIMATE

The Algarve has a Mediterranean climate characterised by long, hot, dry summers with an average temperature range of 20–30°C, and cool, wet winters with an average temperature range of 4–10°C. Its westerly location bordering the Atlantic means that the Algarve is generally wetter and windier (especially in the far west) than the truly Mediterranean regions further east. Most plant growth ceases during the summer, with many plants (so-called ***geophytes***) surviving as dormant underground ***bulbs*** or ***corms*** until new growth is triggered by the first rains in the autumn. The winter rains usually begin in late September or early October, and showers continue into March and April.

The first spring flowers appear in late January, and from mid March to late April the flowering period reaches its peak and the region is covered in a glorious blaze of colour. There is no better place to see this spring flower extravaganza than the Cape St. Vincent peninsula, which is transformed into a natural rock garden of vibrant yellow, blue, pink and white blooms each spring. Year-by-year fluctuations in average rainfall and temperature can be substantial, often leading to shifts in the duration and pattern of the spring flowering period. For example, in 2009 following a cold, wet winter, we noted a delay in the onset of spring flowering and a noticeable shift in the typical composition of the spring flora in the western coastal region, which saw an unusually prolific flowering of geophytes, such as tulips, narcissi, orchids and snowflakes.

As is typical of a Mediterranean climate, frosts are infrequent in the Algarve, allowing the extensive cultivation of tender fruits such as *Citrus* and olives. Indeed, the distribution of the frost-intolerant olive (*Olea europaea*), which roughly tracks the winter 7°C isotherm across Portugal and the rest of the Iberian Peninsula, is a good proxy for the extent of the Mediterranean climate and associated flora within this western 'Mediterranean' extension. Such tender crops are usually grown along the warm southern coastal regions at lower altitude. At higher altitudes in the *serra* (over 600 m), for example in the Monchique region, the climate is much cooler and occasional frosts are possible, so more hardy fruits, such as cherries and plums, are grown along with vegetables, the cultivation of olives and *Citrus* being restricted to the most sheltered valley terraces and lower slopes. Throughout the *serra* and lower hills of the *barrocal*, cork oak, maritime pine and stone pine are extensively cultivated for bark, timber and seeds ('nuts'), respectively. Alongside them, ever-increasing quantities of Australian eucalypts are grown principally to supply wood pulp for the papermaking industry.

GEOLOGY

The interrelated influences of geology, geomorphology and local climate allow the definition of three broad geographical regions of the Algarve: *litoral*, *barrocal*, and *serra*, each with its own characteristic vegetation and flora. The *litoral* consists of the coastal land running along the southern-most strip of the region and up its western coast above the Cape St. Vincent peninsula. It has a mixed geology of sedimentary rocks, alluvia, and sand of relatively recent origin: Quaternary, Tertiary and Cretaceous.

Inland, the *barrocal* is an area of rolling hills, composed principally of limestone (Cretaceous and Jurassic), that increase in altitude upon moving north, where they merge with the mountainous *serra*. The *serra* is composed of older rocks, particularly shale and syenite, mainly Carboniferous in origin. The high (>300 m) *Serra* is predominantly shale with occasional igneous intrusions, most notably the large mass of crystalline syenite at Monchique. The highest peaks of the Algarve, Foia (902 m) and Picota (773 m), are found in this region.

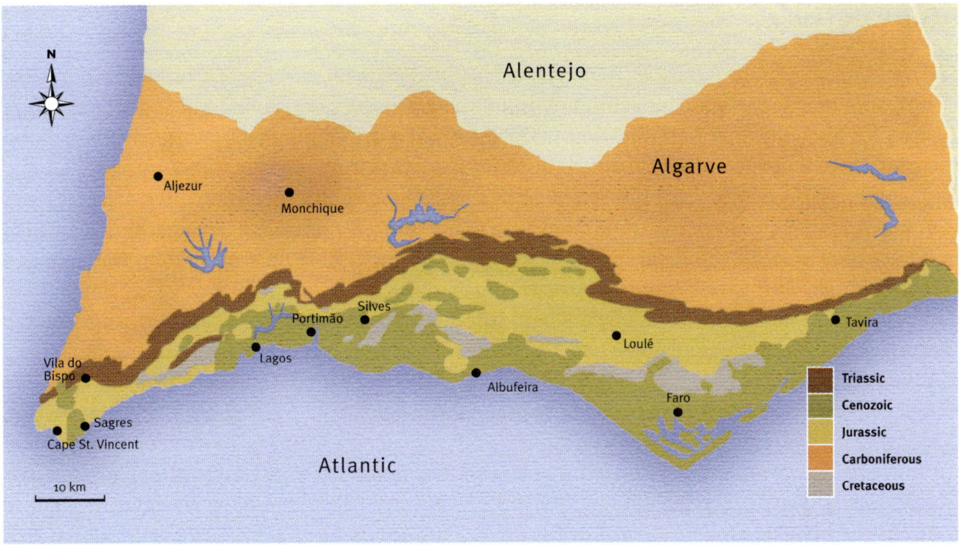

Geological map of the Algarve showing the distinct difference in geology between the north and the south of the region.

Through its influence on soil, particularly soil pH, the geology of these three geographical regions has a significant effect on the composition of their respective floras. Thus, the *litoral* and *barrocal* are dominated by plants that thrive on alkaline limestone soils, such as various species of rock rose (Cistaceae), with lime-intolerant and acid-loving plants being confined mainly to sandy areas and shale outcrops. By contrast, the acid *serra* is dominated by plants that prefer lower pH soils, for instance Ericaceous species such as tree heath (*Erica arborea*) and the strawberry tree (*Arbutus unedo*), fruits from which are used to make the local 'fire water' spirit *medronho*.

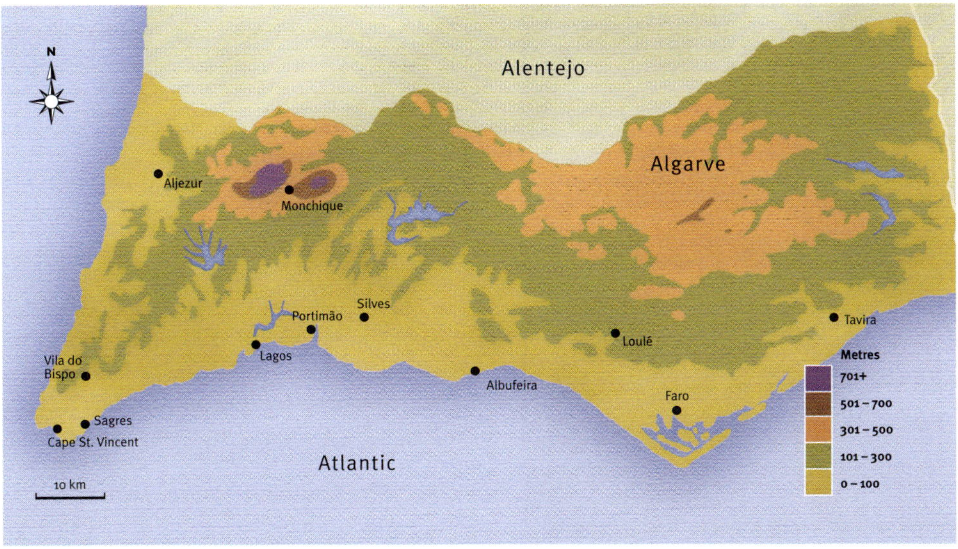

Altitudinal map of the Algarve.

AGRICULTURE

The recent history of the European flora is linked inextricably with the history of humans. Originally, the Algarve landscape was dominated by Holm oak (*Quercus ilex*) forest but most of this was cleared for agriculture in Neolithic times. This has had a profound effect on the landscape and the species composition of plant communities. Today, citrus fruits (most notably oranges, lemons and tangerines) and grapes are important crops in the Algarve, although cultivation of citrus trees is only possible with extensive irrigation. These species rarely ***naturalise*** beyond the boundaries of where they are cultivated, so have not impacted on the native flora away from the confines of agricultural land. Other, more ancient crops (introduced in antiquity) with less demanding cultivation requirements, such as figs, almonds, olives and carob have naturalised, and are now an important part of the flora. All are still cultivated and can be purchased in most local markets, but their cultivation is on a smaller scale than that of the more profitable citrus fruits.

One of the most common ancient plants of cultivation in the region is the carob tree (*Ceratonia siliqua*), which is used for a variety of purposes, including as animal feed (especially for pigs) and as a substitute for cocoa in cheap chocolate. It is understood that the ancient Greeks brought the carob from its native Middle East to Greece and Italy, and that the Moors brought it to North Africa, Spain and Portugal. The ***genus*** *Ceratonia* is the only survivor of an ancient lineage of the pea ***family*** (Leguminosae (also known as Fabaceae)). Interestingly, the term carat, the unit by which the weight of gold is measured, is derived from the Greek *kerátion*, in reference to the ancient practice of weighing gold against carob tree seeds, which display a remarkable consistency in weight and size. Today, Portugal is still one of the world's leading producers of carob.

Carobs (*Ceratonia siliqua*), traditionally cultivated for animal feed and as a cocoa substitute, are commonly naturalised in pastures in the Algarve.

The olive (*Olea europaea*) is another important ancient crop plant of the Algarve and has been a pillar of the Mediterranean agricultural economy for millennia as a source of oil for cooking and burning in lamps. Like that of carob, the ancestry of the olive is poorly understood. The olive has been cultivated in agriculture for so long that truly wild forms no longer exist. Such plants are referred to as 'cultigens' and pose interesting questions for evolutionary biologists and agricultural historians alike.

Another important cultigen that is associated with traditional farming practice is the fig (*Ficus carica*), which is still cultivated on a small scale throughout the Algarve, its fruits being eaten fresh (in summer) or, more usually, dried for consumption all year round. In local markets, 'fig cheese', a much compressed cake of dried figs, is worth sampling, as is the spirit made from figs, *aguardenete de figos*. Figs can often be seen growing alongside almonds (*Prunus dulcis*), another ***deciduous*** tree species that was, and still is, widely cultivated within the region as the fourth element of the traditional four-tree orchard system. Almonds are harvested in late summer, and after drying are processed by crushing machines to release the seeds (nuts) from their hard cases. The nuts are then either eaten whole, used in cakes and confectionery, or crushed to make marzipan; traditional almond-based confectionery especially popular around Easter time. Today, all along the southern coastal region west of Faro, small orchards of carob, olive, fig and almond can be seen in various stages of neglect and reversion to scrub as traditional farming practices have given way to tourism and intensive irrigation-based cultivation of *Citrus* and vines.

Cork Oak (*Quercus suber*) cultivation in the hills near Caldas de Monchique.

On the hills of the *barrocal* and the mountains of the *serra*, one of the most widespread and well-known farming practices is the cultivation of cork oak (*Quercus suber*). Portugal is the world's leading producer of cork, most of which comes from the high plains of the Alentejo, the Algarve's northerly neighbour. The trunk and branches of the cork oak form a thick (fire-resistant) bark, which is harvested for the production of bottle stoppers (corks) and for numerous other applications, ranging from building material and insulation to shuttlecocks. The cork can be harvested sustainably every 9 to 12 years. As the cork-harvesting process is not destructive to the tree, this is a renewable farming practice, and natural stands of cork oak support high levels of biodiversity. The sustainability of cork oak production in the Algarve is therefore integral to the survival of plant species that are specific to the ecosystems of the *barrocal* and *serra*. Cork oak forests are still extensive in these areas, but their continued survival is threatened by the increasing use of plastic 'corks' and screwcaps by the wine industry worldwide (which has lead to a decrease in demand for cork) and by the forestry industry, which can realise quicker and easier profits through the cultivation of fast-growing *Eucalyptus* and pine for the pulp and papermaking industry.

Flora

The Mediterranean Basin contains the richest and most diverse flora in Europe with over 22,000 native species of vascular plants. The Algarve reflects this diversity, having around 1,500 native species of flowering plants. The Algarve's Mediterranean climate supports an extremely rich shrub-dominated vegetation at lower altitudes, and an evergreen oak-dominated woodland at higher elevation. Most of the abundant shrubs are typically woody, often spiny, and have tough waxy leaves, a plant growth form referred to as ***sclerophyllous*** (hard-leaved). The term sclerophyllous refers to the presence of cells that have thick cell walls (sclereids) in the leaves and stems, and these help to retain moisture. Indeed, the sclerophyllous plant form has evolved multiple times in diverse unrelated families in response to the arid conditions experienced by plants in all five Mediterranean climate regions of the world. This means that the vegetation of these geographically disparate regions can look remarkably similar (whether in the Northern Hemisphere or the Southern Hemisphere), a text-book example of 'convergent evolution'. This Mediterranean scrub vegetation is called variously ***maquis*** (France and Mediterranean Basin regions generally), *mattoral* (Spain and Chile), *fynbos* (South Africa), *chaparral* (California) and *mallee* (Australia). *Garrigue* is a term usually used to describe low-lying *maquis* that is associated with dry shallow soils, typical of rocky coastal regions where sea winds might also contribute to the more stunted appearance of the vegetation. In this book, we will use the Portuguese vernacular ***matos*** to describe all sclerophyllous shrub-dominated vegetation, whether *maquis* or *garrigue*.

Stauracanthus genistoides, a common sclerophyllous legume on sand dunes in the Algarve.

Dipcadi serotinum, a frequent geophyte on fixed sand dunes in the region.

The Algarvian *matos* is typically dominated by shrubs of the rockrose (Cistaceae) and pea (Leguminosae) families. In April, the large white flowers of the tall gum rock rose (*Cistus ladanifer*) are an ever-present sight, interspersed on rockier slopes with yellow cushions of *Genista hirsuta*, and on dunes, *Stauracanthus genistoides*.

Most of the non-cultivated parts of the *litoral* and *barrocal* consist of *matos* of 1–3 m in height, but on the cliff tops and plateau of Cape St. Vincent, the vegetation becomes lower (less than 1 m) and more *garrigue*-like, even though the plant species are essentially the same. In spring, the understory of the *matos* is usually rich with flowering geophytes, including narcissi, lilies, tulips, irises and orchids, which having flowered, retreat under ground where they survive the hot dry summer as bulbs, corms or tubers.

Importantly from a botanical perspective, the Algarve province includes the protected South West Alentejo and Cape St. Vincent National Park (*Parque Nacional de Sudoeste Alentejano e Costa Vicentina*), which contains one of the richest floras in Europe. Within this park, the Cape St. Vincent peninsula is probably the most rewarding site for a botanical excursion, especially in spring when the flowers are spectacular. Cape St. Vincent, consisting of a lighthouse, monastery and assorted mobile shops, is the most south-westerly point of Portugal, and lies approximately 6 km from the town and minor resort of Sagres. Here, the cliffs rise to 75 m from the Atlantic and are buffeted by strong gales in winter. Such extreme weather conditions, isolation, and variable geology have together contributed to the evolution of a unique cliff-top flora, which includes a number of ***endemic*** species and ***subspecies*** together with many unique ecotypes. Unfortunately, even within the confines of the park, the flora is threatened by development as tourist apartments spread further inland from the small coastal resorts. The future of many of the rare species in the park is therefore precarious, and will depend on a sensible and sustainable conservation strategy.

ENDEMIC PLANTS OF THE ALGARVE FLORA

There are approximately 50 species of flowering plants that are endemic to the Iberian Peninsula, and at least 12 of these are found only in the Algarve. *The Parque Nacional de Sudoeste Alentejano e Costa Vicentina* is particularly important for Iberian and Algarvian endemic species. For instance, *Silene rothmaleri* occurs on just one or two shale cliff faces in the western Algarve. The plant was discovered in 1945 by the German botanist Rothmaler, and later considered extinct until its rediscovery in 2000[1]. Even before this rediscovery was reported, however, Oxford students had found relatively large numbers of *Silene rothmaleri* growing on the Cape St. Vincent peninsula, and we know of at least three localities where it is still thriving.

[1] Dinter, I. & Greuter, W. (2004). *Silene rothmaleri* (Caryophyllaceae), believed extinct, rediscovered at Cabo de São Vicente (Algarve, Portugal). *Willdenowia* 34: 371–380).

At the extreme end of the Cape St. Vincent peninsula, not far from the lighthouse, the endemic yellow-flowered crucifers *Biscutella vicentina* and *Diplotaxis vicentina* (Brassicaceae) can easily be found growing in sandy pockets among the rocks. *Cistus palhinae* can also be found here. Often viewed as a subspecies of the widespread *Cistus ladanifer* (*C. ladanifer* subsp. *sulcatus*), *C. palhinae* is now largely acknowledged as a distinct species, adding another name to the list of endemics. The lighthouse area is also a particularly good place to see *Astragalus tragacantha* subsp. *vicentinus*, a spiny leguminous shrub whose dome-like forms resemble resting sheep.

Linaria algarviana is a tiny ***annual*** endemic snapdragon with bright purple flowers that can be found along with the extremely rare crucifer *Ionopsidium aucale* on cliff-tops around Aljezur; both of these very rare species are protected by law. Inland from Aljezur, an endemic spurge, *Euphorbia monchiquensis*, can be found in the high-altitude cork oak forests in the Monchique area. This species is subtly different from *Euphorbia characias*, from which it was recognised as distinct only recently. Further east, the very rare endemic *Tuberaria major* (Cistaceae) occurs infrequently in the lower hills of the *barrocal* around Algoz, this being one of the only remaining sites for this dainty, yellow-flowered rock rose.

Silene rothmaleri, an endemic species of the Algarve which occurs on just one or two shale cliff faces in the west of the region.

PLANTS OF PARTICULAR INTEREST

In addition to the rare and endemic species, there are numerous equally fascinating plants that are easier to find. For example, the region is notable for its diversity of parasitic plants, which lacking chlorophyll rob other plants of their nutrients. One of the most spectacular parasitic plants in Europe is *Cistanche phelypaea*, a rare and imposing, bright yellow plant, which parasitises shrubby Amaranthaceae in salt marshes across southern Europe. *Cistanche phelypaea* is relatively common in the Algarve and in some places it can be found in great stands, for instance in the salt marshes around Faro, Lagos and Alvor. Another even rarer parasitic species that grows in these areas is *Cynomorium coccineum*; this is such a derived ***parasite*** that it barely resembles a plant at all and looks more like a fungus. Tourism-driven development has drastically reduced the abundance of this species, and it is difficult to find, even at Alvor, one of its previous strongholds, but it still occurs locally on cliff tops near Portimão.

The Algarve is a renowned site for the rare carnivorous plant *Drosophyllum lusitanicum*. This species is unusual among carnivorous plants because it grows on dry alkaline soils rather than in acidic water-logged marshes. Like the sundews (*Drosera* spp.), *Drosophyllum* produces ***rosettes*** of sticky leaves, which are covered in mucilage that traps small insects. Unfortunate insects become ensnared on the sticky leaves and eventually become asphyxiated and digested by enzymes exuded by the leaf. *Drosophyllum* grows only in Spain, Morocco and Portugal and is very uncommon. It was once more plentiful in the Algarve, especially in the *barrocal* near Algoz, but has suffered at the hands of plant collectors and urban development. Nevertheless, ongoing conservation efforts are using tissue-culture-based methods to propagate this rare and unusual plant, so its future in the region appears promising.

The Algarve is also an ideal place to see orchids. Orchidaceae is one of the largest families of flowering plants, with over 20,000 species, most of which are tropical and sub-tropical in their distribution. However, Mediterranean climate regions are also particularly rich in orchids and contain some of the most unusual forms in terms of their pollination biology. Bee orchids (*Ophrys* spp.) are common in the Algarve and can be found growing in abundance in 'weedy' and neglected orchards of olive, fig, carob and almond. Pollination of bee orchids relies on sexual deception: the flowers of the various species of *Ophrys* mimic the females of particular species of bee, wasp or fly, both visually and also chemically by producing a sex-pheromone mimic. Male insects are attracted to the flowers by their scent and appearance and attempt to mate with them, a process called ***pseudocopulation***. During pseudocopulation, discrete packets of pollen (***pollinia***) become attached to the male insects, and these are then inadvertently transferred to the ***stigma*** of the next flower they visit, thereby effecting pollination and subsequent fertilisation. Seven species of *Ophrys* are locally abundant in the Algarve, the most common of these being the yellow bee orchid (*O. lutea*), the bumblebee orchid (*O. bombyliflora*) and the mirror orchid (*O. speculum*). Tongue orchids (*Serapias* spp.) are also quite common throughout the *litoral* and *barrocal* and these possess another intriguing mode of pollination. All species of *Serapias* bear small tube-like flowers, which insects often use as a refuge from rain or as an overnight resting place. While the insects are sheltering within these flowers, sticky pollinia become attached to their bodies. Pollinia are then transported to other flowers in which the insects subsequently seek refuge.

Cistanche phelypaea, an imposing yellow parasite of halophytic shrubs on dunes and salt marshes, photographed here at Alvor.

Cynomorium coccineum growing on cliff-tops near Castelos. This poorly understood parasite is very rare and is now restricted to just a few unstable cliff-tops in the region.

Drosophyllum lusitanicum, a very rare carnivorous plant that has declined markedly in the Algarve and could face extinction in the region. Photograph by Finn Rasmussen.

The mirror orchid (*Ophrys speculum*) is one of several orchid species in the region that exploit male wasp pollinators by duping them into attempting to mate with their flowers, and in so doing, bring about cross-pollination.

A wasp seeking refuge in the flower of a tongue orchid (*Serapias lingua*) during a rainy spell, with pollinia adhered to its head.

Italian arum (*Arum italicum*) lures midges with the false promise of food, then imprisons them and showers them with pollen before releasing them to cross-pollinate the next inflorescence they visit.

Among plants of the Algarve, strange and bizarre mechanisms of pollination are not confined to the orchids. The inflorescences of species of Dutchman's pipe (*Aristolochia* spp.) and Italian arum (*Arum italicum*) function as traps for pollinating insects and offer no reward for pollination services. *Arum italicum*, which is common in damp woodland in the *serra*, produces an ***inflorescence*** with a yellow spike (***spadix***), which generates heat and becomes warmer than the surrounding environment, a process called thermogenesis. This, coupled with a smell resembling cow dung, attracts midges, which crawl into a chamber of the inflorescence sheath (the ***spathe***), and become trapped by a ring of downward-pointing spines. Once trapped, the midges are retained within the trap for approximately 24 hours, so that any pollen they may be carrying from another *Arum* becomes deposited on the stigmas. After this time, the midges are showered with pollen before being released as the spines wither, thereby ensuring that the next inflorescence they visit will be cross-pollinated.

Wild flower habitats

COASTAL (LITORAL) PLANT COMMUNITIES

Much of the southern *litoral* is developed for tourism or threatened by grazing, but its western end (west of Lagos), within the *Parque Nacional de Sudoeste Alentejano e Costa Vicentina*, is much less developed, and northwards from Cape St. Vincent the coast remains largely unspoiled. Owing to the contrasting aspects and microclimates of the southern and western *litoral*, the plant communities in these two regions are very different. Much of the south coast is bordered by sheltered coves with limestone cliffs and low coastal *matos*, whereas the west coast is predominantly flanked by long, sandy beaches and extensive dune systems. These relatively undeveloped regions of *litoral* support some very rich maritime plant communities, and are arguably some of the best coastlines for botanising in Europe.

Coastal habitats are typically hostile environments where plants must jostle against the elements to survive. All coastal habitats are subject to high salinity, and the plants that grow there have evolved to tolerate the salty conditions. Such plants are called ***halophytes*** because they possess specific adaptations that allow them to survive at salinity levels that would be toxic to other plants. Close to the sea, halophytic plants can cope with high levels of salt and even submersion in sea water. Indeed, sea grasses (such as *Zostera* spp.) are unusual among higher plants in that they live permanently immersed in sea water. Further inland, plants tolerate salt-laden sea spray and fierce winds, especially along the cliff-tops of Cape St. Vincent. Many

Sheep and goat grazing has a profound effect on species composition in the Algarve by driving negative shifts in the abundance of palatable species and an increase in the abundance of unpalatable species, for example here: *Ononis natrix*.

halophytes have a bluish (***glaucous***) appearance resulting from a waxy coating on their leaves that protects them from sun and salt damage; others have hairs and special salt-secreting *glands*. Particularly good places to see halophytes are the cliffs of the southern *litoral* where the attractive, yellow-flowered *Asteriscus maritimus* is common, among the rocky outcrops near the lighthouse on Cape St. Vincent, and on the shale cliffs of the west coast, which support geophyte communities.

The sand dune plant communities in the Algarve are typical of those throughout Europe. The fore-dunes are dominated by marram grass (*Ammophila arenaria*), which plays a vital role in stabilising shifting sands by virtue of its extensive mesh-like root system. Other species that are common on sand dunes include the sea daffodil (*Pancratium maritimum*), which is a bulbous ***xerophyte*** that carries showy white flowers in mid summer, sea spurge (*Euphorbia paralias*), *Lotus creticus* and *Crucianella maritima*. Minute micro-annuals grow in abundance on fixed dunes, these include the four-leaved allseed (*Polycarpon tetraphyllum*), which completes its life cycle in just a few weeks. Across the dunes, from the sea shore inland, there is a succession of increasing plant diversity and habitat complexity as the sand becomes more stabilised and nutrient-rich, finally grading into a typical *matos* plant community.

Probably the best place to see and appreciate the coastal plants of the Algarve is on the cliff-tops at Cape St. Vincent, an area which is best seen in spring (mid March to early May). The species composition on the cliffs is complex but it is possible to identify distinct vegetation zones, owing in part to the effect of sand deposition, the result of the tsunami that followed the earthquake of 1755 which virtually destroyed Lisbon. This has given rise to a seemingly incongruous mosaic of isolated sand dune plant communities alongside more typical coastal *matos* communities on these high cliffs.

The German botanist Rothmaler identified and mapped numerous distinct vegetation zones on Cape St. Vincent between 1938 and 1940. Towards the lighthouse, some of these zones are

Exposed sea cliffs are exposed to gales and salt spray, and support communities of halophytic plants such as *Asteriscus maritimus* photographed here at Boca do Rio.

Shale cliffs are a common geological feature along the west coast of the Algarve, and are important maritime habitats that support geophytes such as rare *Narcissus* species.

Ammophila-dominated sand dunes at Amoreira, supporting populations of *Lotus creticus* (bottom left) and *Pancratium maritimum* (top centre).

Distinct plant communities on Cape St. Vincent: *Corema*-dominated dunes (top) and communities dominated by *Stipa gigantea* and *Halimium calycinum* (bottom).

easily recognised, particularly a zone of sand dune communities dominated by acidophilic xerophytes, such as the heather-like *Corema alba* (Empetraceae), growing with Spanish oat-grass (*Stipa gigantea*) and yellow-flowered *Halimium halimifolium*. Closer to the cliff edges is a zone dominated by *Cistus palhinhae* and stunted junipers (*Juniperus phoenicea*). Between the cliff-top *Cistus–Juniper* zone and the acidic sand dune zone is a highly heterogeneous zone with the greatest diversity of species, most notably aromatic herbs, such as thyme (*Thymus camphoratus*), lavender (*Lavandula stoechus*) and rosemary (*Rosmarinus officinalis*), interspersed with the spiny legumes *Stauracanthus genistoides* and *Astragalus tragacantha* subsp. *vincentina* and the colourfully-flowered *Halimium calycinum*, *Lithodora diffusa*, *Antirrhinum majus* and *Anagallis monellii*. In early spring, geophytes such as the little rust-coloured 'bluebell' *Dipcadi serotinum*, *Fritillaria lusitanica* and *Allium subvillosum* are common, along with the sombre bee orchid (*Ophrys fusca*) and the two-leaved gennaria (*Gennaria diphylla*). In April, the yellow flowers of the endemic crucifers *Diplotaxis vicentina* and *Biscutella vicentina* are easily spotted. Other endemics to look out for include *Ionopsidium aucale* and *Linaria algarviana*. In addition to the ever-present threat posed by increasing levels of tourism and associated development, parts of the Cape are also threatened by the smothering effect of the invasive ***alien*** Hottentot fig (*Carpobrotus edulis*), which was introduced to Europe from South Africa.

SALT MARSH PLANT COMMUNITIES

The salt marshes around the Algarve are of particular interest to botanists and ornithologists alike. The most extensive salt marsh systems lie to the east of the region where the mud flats of the Ria Formosa nature reserve extend from Faro in the west to Tavira in the east, covering an area of approximately 3,000 ha. Here, bordering the Guadiana River, are the salt pans of Castro Marim, as well as lagoons, islands, dune systems and salt marshes. The area is well known to ornithologists for its diversity of bird species. To the west of Faro there are other, albeit smaller, interesting salt marsh plant communities, for example those near Portimão, Lagos, Alvor and the mouth of Ribeira de Aljezur by the dune slacks at Amoreira. With the exception of the Ribeira de Aljezur, these habitats are all at least partially degraded by development.

Salt marshes are inhabited solely by halophytic plant communities, which are often submerged by tidal waters. In European salt marshes, plant communities are dominated by species of the family Amaranthaceae (formerly Chenopodiaceae), which are characterised by their ability to withstand saline conditions. Particularly common are the shrubby species *Arthrocnemum perenne*, *Atriplex halimus* and *Salicornia europaea*, which are characteristically bluish and succulent. These habitats are also home to two rare species of parasitic plant, *Cistanche phelypaea* and *Cynomorium coccineum* (see 'Plants of Particular Interest').

MATOS

Matos is composed of sclerophyllous plants that are adapted to survive in dry hot conditions. It is the most common vegetation type in the Algarve and characterises the landscape of the region at low to mid altitude from the *litoral* into the *barrocal* (see 'Flora'). Most *matos* vegetation consists of woody shrubs of 1–3 m in height, with gum rock rose (*Cistus ladanifer*) usually the dominant species. This attractive, bushy rock rose is well adapted to the arid conditions, having a sticky residue ('gum') on the surface of its leaves that protects it from ***desiccation*** and also serves to reflect sunlight. This resin (ladanum), which is extracted by boiling twigs and skimming off the non-aqueous surface layer, has been used in soaps and perfumery since ancient times.

Matos of the coastal *litoral* is usually fairly low in height (shorter than 1 m) owing to the combined effects of thin, dry, rocky soils and coastal winds. On the wind-swept plateau of Cape St. Vincent, *Cistus palhinhae*, a close dwarf relative of *Cistus ladanifer* (possibly an ecotype)

Low coastal *matos* (*garrigue*) on the cliff-tops of Cape St. Vincent, dominated by the endemic, white-flowered *Cistus palhinhae*.

Typical middle *matos* photographed near Salema dominated by *Cistus albidus*, *Cistus ladanifer*, *Ficus carica* (centre right) and mastic (*Pistacia lentiscus*) (right).

dominates the low *matos*. Other common species of this *garrigue*-like *matos* (see 'Flora') include woody shrubs: juniper (*Juniperus phoenicea*), mastic (*Pistacea lentiscus*), daphne (*Daphne gnidium*), two species of buckthorn (*Rhamnus alaternus* and *Rhamnus lycioides*), and the olive-like *Phillyrea angustifolia*, all of which assume a growth form that is stunted and dwarfed relative to their normal appearance in more sheltered inland regions.

Away from the coast, these woody *matos* shrubs can be quite tall (up to 3 m or above) and dense, with *Cistus ladanifer*, *Pistacia lentiscus*, *Phillyrea angustifolia*, *Juniperus phoenicea*, *Daphne gnidium*, *Rhamnus alaternus* and *Rhamnus lycioides* making up the bulk of the vegetation. Smaller shrubs include Spanish lavender (*Lavandula stoechas*), grey-leaved rock rose (*C. albidus*, with pink flowers), narrow-leaved rock rose (*C. monspeliensis*, with small white flowers and non-sticky leaves), and the hairy legume *Genista hirsuta*. Occasional neglected figs, almonds, olives and carobs can be seen among the vegetation, indicating that the land now covered by *matos* was once under traditional orchard cultivation.

In most areas of *matos*, it is possible to find the curious yellow parasite *Cytinus hypocistis* growing beneath its usual host *C. monspeliensis*. More locally, its sister species, the red and white *Cytinus ruber*, parasitises the roots of *Cistus albidus*. Common geophytes of the *matos* include *Tulipa sylvestris*, *Gladiolus italicus*, *Asphodelus ramosus* and *Narcissus bulbocodium*, as well as the sea squill (*Urginea maritima*), which has very large bulbs and forms distinctive inflorescences up to 1.5 m tall in late summer. Extracts from the bulbs of sea squill were once used as a form of rat poison. Orchids too are common in the *matos*, most notably the Italian man orchid (*Orchis italica*), green-winged orchid (*Anacamptis morio*) and assorted bee orchids (*Ophrys* spp.)

Interestingly, in the barer areas of *matos* and in recently abandoned orchards reverting to *matos*, berry-forming shrubs can be seen in clustered aggregates. These 'islands' of different berry-producing species, such as juniper, mastic, buckthorn and daphne, are the result of the simultaneous dispersal of seeds that have been ingested by birds. A similar phenomenon can be seen in neglected orchards of figs, where seeds from the droppings of perching birds, especially blackcaps (*Sylvia atricapilla*), have germinated to form a circle of different berry-forming shrubs around the trees. Within these shrubby aggregates, the hemi-parasitic *Osyris quadripartita* (which also produces berries) can usually be found parasitising the roots of the other species with which its seeds were co-dispersed.

PINE FORESTS

Pine forests are common throughout the region but especially in the *barrocal* and *serra*. The stone or umbrella pine (*Pinus pinea*) is the source of the edible pine 'nuts' (*pignons*) and also of timber, and is commonly planted in regimented plantations in the *barrocal*. The closely related Aleppo pine (*Pinus halepensis*) forms extensive forests on higher ground, where it is grown principally for timber and wood pulp. Within the *litoral* there are fewer pine forests, and these consist mainly of maritime pine (*Pinus pinaster*), which is grown for timber and turpentine. Pine forests are usually less biodiverse than the *matos*, but are of interest in their own right. Species that are commonly associated with pine forests include acidophilic shrubby legumes, such as *Stauracanthus* spp., *Ulex* spp., *Cytisus* spp. and *Chamaespartium* spp., heathers (*Erica* spp.) and *Cistus crispus*. A number of orchid species inhabit pine forests, including the tongue orchids *Serapias vomeracea* and *Serapias cordigera*, helleborine (*Epipactis tremolsii* subsp. *lusitanica*) and locally the unusual violet limodore (*Limodorum abortivum*), which parasitises the roots of pine trees via a fungal symbiont in a process called ***mycoheterotrophy***.

Stone pines (*Pinus pinea*) in the Western Algarve.

OAK FORESTS

At higher altitudes in the *serra*, the climate is cooler and annual rainfall greater. This gives rise to sub-montane oak forest plant communities that are markedly different from the coastal and hillside plant communities at lower altitude. Post-glacial warming of the continent once forced plants that were adapted to cooler regions to migrate into habitats at higher altitude, so montane forests in the Mediterranean share many of the species that are common in the temperate forests of Northern Europe. Most of the original forests of the Algarve have been felled and replanted with *Eucalyptus* spp. and various species of pine and fir (*Abies* spp.). However, extensive native cork oak (*Quercus suber*) forests are still widespread and important both economically (see 'Agriculture') and ecologically because they support a rich biodiversity of plants, fungi and animals. Common plant species in cork oak forests include the strawberry tree (*Arbutus unedo*), tree heath (*Erica arborea*), *Vibernum tinus*, green lavender (*Lavandula viridis*), broom (*Cytisus scoparius*) and butcher's broom (*Ruscus aculeatus*). Locally, *Rhododendron ponticum* (native to the Himalayas) and the sweet chestnut (*Castanea sativa*) are also common, the latter being important economically as a source of fence posts and building materials. Other plants of interest in these forests include birthwort (*Aristolochia longa*), with its peculiar pitcher-like flowers that imprison pollinating insects, and the beautiful red-flowered peony *Paeonia broteroi*, which can be found in cork forests around the picturesque *serra* town of Monchique. Here, on the roadside near Picota, it is also possible to see the last remaining specimens of a majestic deciduous oak, *Quercus canariensis*; this was once a prominent element of the region's oak forests before Neolithic settlers cleared these trees. Another rarity of this area is *Euphorbia monchiquensis*, which grows locally among mossy boulders. Many alien exotics have been planted in and around upland villages in the *serra* and some, such as the arum lily (*Zantedeschia aethiopica*) native to South Africa, have escaped to the surrounding hillside.

DISTURBED HABITATS

Any environment where the soil has been turned, including roadsides, paths, loose cliffs, arable land, demolition sites and gardens, qualifies as a 'disturbed' habitat. Roadsides and derelict urban areas are often surprisingly rich in native annuals, opportunistic short-lived ***perennials*** and alien exotics. All these plants are adapted to disturbed and changing environments. Typically, they display high rates of reproduction, rapid growth and often tolerance of dry conditions. Disturbed sites tend to be of ephemeral interest to the botanist; they are best explored in spring because by June they have usually dried up and become dead and uninteresting.

Disturbed coastal sites are particularly rich in native annuals. At least 11 species of spurrey (*Spergularia* spp.) occur in the region, and many of these grow on disturbed sandy ground, such as car parks or near buildings by the coast. Disturbed sites are also fertile ground for annuals in the daisy family (Asteraceae), among the most common of which are *Chrysanthemum coronarium* and *Ditrichia viscosa*. Spikes of parasitic broomrapes (*Orobanche* spp.) are locally common on disturbed ground, where their minute seeds have easy access to the roots of their specific host plants.

The extensive development of the Algarve has also led to the widespread introduction of alien exotics, which have had a profound effect on the local landscape. For example, the century plant (*Agave americana*) with its tree-like inflorescences, which persist for years after they have bloomed, often line the roadsides and can be seen from miles away. The Bermuda buttercup (*Oxalis pes-caprae*) from South Africa is another widespread alien, which forms carpets of yellow flowers along roadsides throughout April and May. Similarly, the Australian mimosas, such as *Acacia retinodes* with cascading bunches of yellow flowers, which were originally planted as hedges and wind breaks, dominate roadsides in spring. Another alien hedgerow weed that has spread widely in the Algarve over past few years is the pretty pink- and yellow-flowered *Lantana camara*, another South African exotic. Some alien species pose a serious threat to native biota, for example the Hottentot fig (*Carpobrotus edulis*) often takes a hold in disturbed sites, where it smothers and suffocates the native plants.

A recently disturbed roadside colonised by *Chrysanthemum coronarium*, a common weed of disturbed ground.

Identifying and classifying wild flowers

WILD FLOWER CLASSIFICATION

For centuries, classifications of living organisms relied on the use of morphological characters to group related species into genera, genera into families, and families into orders. Such classifications were often based on the wisdom of expert individuals or small groups of experts. Thus, they were built around informed opinions rather than a rigorously (statistically) tested standard taxonomic data 'currency'. Since the 1990s, however, advances in the technologies of genetic analysis and computational statistics have resulted in a revolutionary shift towards classifications that are based on genetic characters revealed by the sequences of DNA, which have provided a universal taxonomic data currency. This DNA-based method of classification, called molecular phylogenetics or molecular systematics, is now used in classifications of all living organisms, including plants, and generally produces more 'natural' classifications that reflect more directly the evolutionary (genealogical) relatedness of different groups of organisms. This is because shared gene sequences can be used to identify 'monophyletic groups' — natural groups that contain all the descendants of a common ancestor. In many cases, molecular phylogenetics has confirmed (and thus strengthened) previous morphology-based groupings, whereas for other groups it has identified previously hidden genealogical relationships that were impossible to infer from morphological characters alone. (See: www.mobot.org/MOBOT/research/APweb/ for a fuller account of molecular phylogenetics.)

Today, the most widely used system for classifying flowering plants is DNA-based and is the product of the 'Angiosperm Phylogeny Group' (APG), which produced its first major classification of angiosperms in 1998, with subsequent revisions in 2003 (APGII) and 2009 (APGIII). We therefore follow the APG system of classification in this Flora. For this reason, the assignment of particular species to genera and families in this book will sometimes differ from that seen in older traditional Floras and texts, notably *Flora Europaea*, *Flora Iberica*, and the widely used Mediterranean floras authored by Oleg Polunin and colleagues: *Flowers of the Mediterranean* (Polunin & Huxley, published by Chatto and Windus) and *Flowers of South-West Europe* (Polunin & Smythies, published by Oxford University Press). For example, many species formerly placed in the family Scrophulariaceae have now been assigned to other families, most notably the Plantaginaceae (plantain family), which now contains the genera *Antirrhinum* (snapdragons) and *Linaria* (toadflaxes). More recently, the asparagus family (Asparagaceae) has been greatly expanded to include, among others, the former families Agavaceae, Ruscaceae and Hyacinthaceae.

Even though classifications may now be DNA-based, species descriptions, which are essential for field identification, must by their nature be morphology-based, at least until we have the appropriate technological devices to read species-specific DNA 'barcodes' in the field. Until such technology arrives (and arrive it will), we must rely upon morphological descriptions and illustrations such as those presented in this Flora.

WILD FLOWER MORPHOLOGY

An understanding of the basic anatomy of plants, especially of flowers and inflorescences, is essential for wild flower identification, because their form will frequently indicate or suggest the family to which a species belongs. For example, an ***umbel*** is an inflorescence characteristic of the carrot family (Apiaceae, formerly the Umbelliferae), whereas a ***capitulum*** is an inflorescence typical of the daisy family (Asteraceae). The symmetry of individual flowers is also important: bilaterally symmetrical (***zygomorphic***) flowers are characteristic of the pea family (Leguminosae-Fabaceae), mint family (Lamiaceae) and plantain family (Plantaginaceae), whereas radially symmetrical (***actinomorphic***) flowers are characteristic of, for example, the carnation family (Caryophyllaceae), rock rose family (Cistaceae), and mustard family (Brassicaceae). Care should always be taken when observing the form of flowers because subtle morphological details can

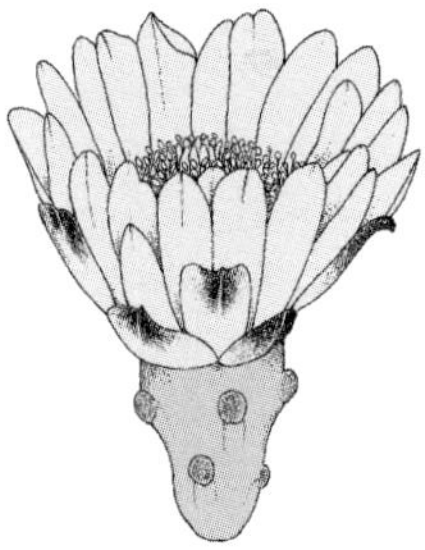

Solitary flower
A flower that is borne singly, not part of an inflorescence

Spike
Stalkless flowers attached to a stem

Raceme
Stalked flowers attached to a stem

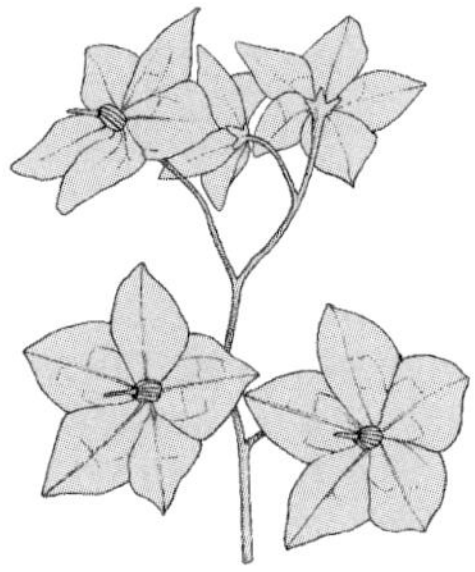

Cyme
An asymmetrical, rather flat-topped inflorescence in which the central flower normally opens first

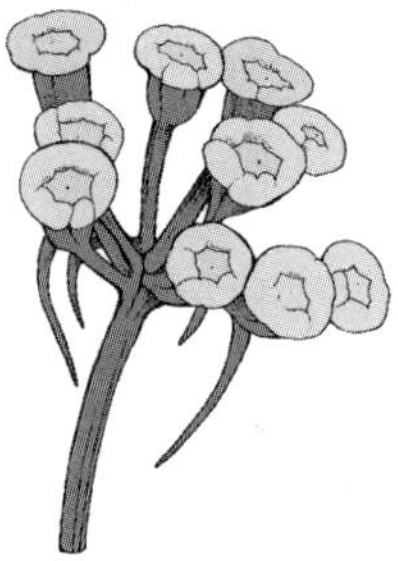

Umbel
Flowers with stalks, often of different lengths, forming an umbrella-shaped inflorescence

Corymb
A rather flat-topped inflorescence with flowers on stalks of different lengths

Panicle
Stalked flowers borne from repeatedly branching stems

Capitulum
Many tightly packed tiny flowers, often forming a central disc

often be overlooked, especially in very small flowers; for example, the capitulum of a daisy is not a single actinomorphic flower but rather an inflorescence consisting of many minute flowers (***florets***), which may be zygomorphic (e.g. in *Taraxacum* spp.), actinomorphic (e.g. in *Cotula* spp.), or a mixture of the two (e.g. in *Bellis* spp. and *Chrysanthemum* spp.). For this reason, the use of a hand lens is recommended.

Actinomorphic flowers

Actinomorphic flowers are radially symmetrical, meaning the flower has multiple planes of symmetry when viewed from above. They are typically pollinated by generalist insect pollinators. Many groups of flowering plants bear actinomorphic flowers, classical examples being buttercups (*Ranunculus* spp.), poppies (*Papaver* spp.) and lilies (*Lilium* spp.).

Mallow family (Malvaceae)

Mustard family (Brassicaceae)

Pink family (Caryophyllaceae)

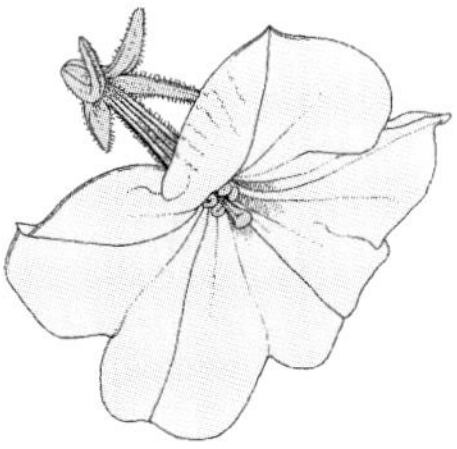

Potato family (Solanaceae)

Zygomorphic flowers

Zygomorphic flowers are bilaterally symmetrical, meaning they have a single plane of symmetry that divides the flower into two mirror-image halves. Zygomorphic flowers are typically fertilised by specialist pollinators and some families have evolved highly complex flowers associated with particular types of pollinator. The most specialised zygomorphic flowers are seen in the orchids (Orchidaceae). Other classical examples include the snapdragons (*Antirrhinum* spp.), foxgloves (*Digitalis* spp.) and members of the pea family (Leguminosae).

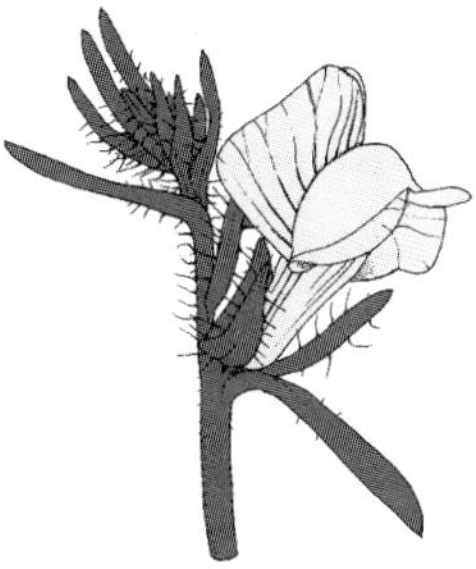

Snapdragon family (Plantaginaceae)

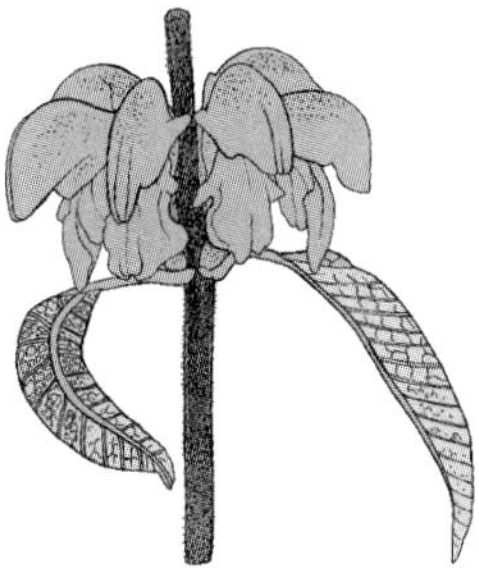

Mint family (Lamiaceae)

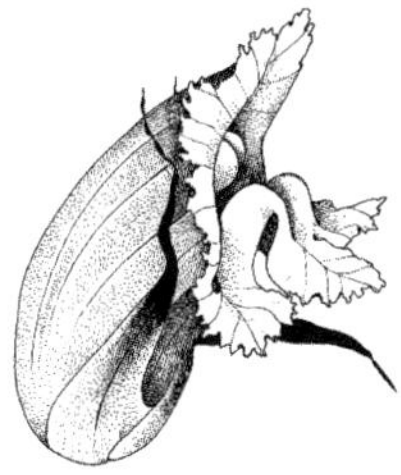

Broomrape family (Orobanchaceae)

Pea family (Leguminosae)

How to use this book

The bulk of the plants described in this Flora are angiosperms (flowering seed plants), but as in most Floras we also include the most common gymnosperms (woody non-flowering seed plants, such as conifers), as they often form a major component of the vegetation. In accordance with the APG system, we divide the flowering plants of the region into three principal groups: basal angiosperms (including the Magnoliids, about 5% of species), monocots (about 25% of species), and eudicots (about 70% of species worldwide). The latter two groups equate to the former subclasses monocotyledons and dicotyledons, respectively, but because the former dicotyledons was shown not to be monophyletic by APG, we now recognise an assemblage of ancient former dicots (basal angiosperms) and the monophyletic group eudicots ('true' dicots). The basal angiosperms consist of a number of small monophyletic lineages, representing the most ancient ('early divergent') lineages of flowering plants; these include the water lilies, magnolias, laurels, and many other species not native to Europe. In the Algarve, native species in the Piperaceae and Aristolochiaceae families are basal angiosperms within the Magnoliid complex. The monocots include the grasses, lilies, orchids and other bulbous geophytes, and are usually easily recognisable by their parallel-veined leaves, generally ***herbaceous*** habit, and flower parts arranged in multiples of three. The eudicots, by contrast, can be woody or herbaceous, usually have net-veined leaves, and have flower parts in multiples of two, five or seven.

The sequence of species descriptions follows APG II, which groups related families into orders (ending in -ales). Thus the fig (genus *Ficus*, specific epithet *carica*) is in the family Moraceae, which is part of the order Rosales. The selection of species described is based largely on the observations made by the authors while studying the flora of the region, supplemented by species described in other texts. Inevitably, the coverage of species will be biased towards the areas we have studied in most detail, but we have tried to include among the descriptions those species most likely to be encountered and those of particular interest (for example, rarities, endemics and species with unusual pollination mechanisms). With approximately 1,000 ***taxa*** described, this is the most comprehensive Flora of the region written to date.

DESCRIPTIONS

The species descriptions provide detailed information about the appearance of each plant. Where species have a common name, this has been included. For some species, new names have been given recently or more than one name is commonly used in the literature. For these species, the most common synonym has been included (for example, *Silene uniflora*).

As this Flora is intended to be used as a portable fieldguide, we have kept the descriptions brief, concentrating only on those characteristics essential for identification. Species that are similar have been highlighted to avoid confusion between taxa. The size dimensions are approximate, and may vary under different environmental conditions, but should allow a general comparison to be made between species. The descriptions include details of leaves, flowers and fruits; it should be noted that these might not all be present for a particular species at a given point in time. The habitat described is often useful for identification as many species are restricted to a particular habitat, for example coastal environments. Other species are more cosmopolitan and occur in a range of habitats. The flowering time is generally not described. This is because most of the species described will be in flower in the region between March and mid May. Exact flowering times vary from year to year depending on the weather. For those species that do not flower in spring, the flowering time is given with the description. The distribution of a given species within the region is an indication only and should be treated with care; there is limited literature on the geographical distribution of plants in the Algarve, so the information given in descriptions is based largely on the observations made by the authors.

We have endeavoured to avoid excessive use of technical terms to allow for easy use by those unfamiliar with botanical terminology. Definitions for commonly used botanical terms (which are highlighted in bold and italic text at first mention) can be found in the 'Glossary' on p. 245. Where it has been necessary to use an uncommon term as part of a description, a definition is included in the description.

ILLUSTRATIONS

Black and white illustrations are provided alongside most of the descriptions as an aid to identification, highlighting particular features of the plant that are useful for identification. To save space, not all of the species described have illustrations. The illustrations are particularly useful for groups of closely related and superficially similar species; for such groups, illustrations are included for as many taxa as possible.

PHOTOGRAPHS

Owing to the comprehensive coverage of this Flora, it has been impossible to illustrate every species with a photograph. Therefore, photographs have been selected to illustrate common species from all of the main families and genera, and in particular those species that are most likely to be encountered or are of particular interest. By and large, the photographs depict the flowering stage, which is generally the most conspicuous and most useful for identification purposes. Where the habit (growth form) of the plant is more useful, for example for many tree species, this has been depicted instead of the flower.

Species descriptions

GYMNOSPERMS

Non-flowering seed plants, including the conifers, in which the **ovules** are 'naked', rather than enclosed within an **ovary**.

Order: **PINALES (including GNETALES)**

PINACEAE

Evergreen trees or shrubs with whorled branches and needle-like leaves, flowers borne in cones; male cones on young shoots with copious pollen, female cones borne on the same tree and more substantial, with woody scales, and taking years to mature. Many species are cultivated in the Algarve.

P. pinea

Pinus pinea **STONE PINE, UMBRELLA PINE** A large, distinctive tree to 30 m, **parasol-shaped** when mature. Bark grey-brown with peeling red patches. Leaves green, 10–20 cm long, to 2 mm wide, with minute, forward-pointing **teeth** along the margins. Cones broadly **ovate**-spherical, shiny, red-brown with large, unwinged or **scarcely winged seeds**. Common on hills and in coastal habitats, often in large stands, frequently planted for the edible seeds. ***Pinus halepensis*** is a similar tree but with greyish twigs, a more rounded crown and **seeds with well-developed wings**. Local on heaths.

P. halepensis

Pinus sylvestris **SCOTS PINE** Tall tree to 40 m. Bark dark brown below, pale red and flaking above. **Leaves small**, to 70 mm long, 3 mm wide, distinctly stalked, and in pairs with a persistent grey basal sheath; **glaucous** and twisted. Cone more or less **globose**, acute, deflexed and yellowish-brown.

Pinus pinaster **MARITIME PINE** A tall, pyramidal tree to 40 m, often flat-topped and stunted in coastal stations. Bark blackish, becoming fissured. **Leaves in pairs,** dark green, **long (to 25 cm)**, **spine-tipped**. Female cone ovoid, in clusters of 2–8, light brown and shiny with spiny scales. Common on maritime sands and widely cultivated for timber, soil conservation and the production of turpentine.

P. sylvestris

P. pinaster

1. *Pinus pinea*
2. *Pinus halepensis*
3. *Pinus sylvestris*
4. *Pinus pinaster*
5. *Juniperus phoenicea*
6. *Juniperus phoenicea* fruits

CUPRESSACEAE

Trees or shrubs with resin, leaves opposite or in whorls, scale-like or needle-like. Fruit a cone, often berry-like.

Juniperus phoenicea **PHOENICIAN JUNIPER** A tall shrub to 8 m, with grey-brown bark peeling in narrow strips. Leaves to 15 mm long, smaller on older branches and scale-like with **membranous** margins, and appressed to the stem. Female cones spherical, red-brown when mature. Very common on the *matos* and on dry, stony ground.

1
2
4
3
5
6

Juniperus oxycedrus* subsp. *transtagna PRICKLY JUNIPER A greyish **dioecious** shrub to 8 m (usually less) with sharply pointed needle-like **leaves less than 1 mm wide, with 2 pale bands above**, in whorls of 3. Cones 8–10 mm in diameter, ripening red-brown. Local in dune systems; frequent in Carrapateira.

Cupressus sempervirens CYPRESS A distinctively narrowly columnar tree to 30 m tall. Leaves dark green, scale-like to 1 mm long, closely overlapping. Female cones spherical, to 40 mm across and yellowish. Widely planted in towns and gardens in the Algarve.

ARAUCARIACEAE

An ancient family of evergreen, coniferous trees from the Southern Hemisphere with columnar trunks and, usually, needle-like leaves. Trees produce either male or female cones separately (dioecious) or together on the same tree (monoecious). Seeds usually borne singly on the upper surface of scales (in contrast to true pines in which seeds are borne in pairs).

A. heterophylla

Araucaria heterophylla A tall, erect tree to 60 m, with horizontal spreading branches arranged somewhat symmetrically in whorls around the trunk. Cones squat and globose. Native to Norfolk Island in the Pacific; widely planted in the Algarve.

EPHEDRACEAE

Small, dioecious shrubs with rush or broom-like stems, small opposite or whorled leaves and berry-like fruits.

E. distachya

E. fragilis

Ephedra distachya A **low, scrambling shrub (to 1 m)** with erect stems from creeping **rhizomes**, forming a thicket. **Stems rigid** and broom-like, leafless except for scale-like leaves to 2 mm at joints. Male and female flowers separate, greenish-yellow and small; male flowers in clusters, the female in pairs. Fruit globose, berry-like and reddish, the seed protruding. Local on dry rocky slopes, sea cliffs, and river banks.

Ephedra fragilis JOINT PINE Similar to *E. distachya* but much **taller**, climbing or prostrate to 5 m with disordered, **flexible** but fragile stems that break easily at the nodes. Leaves scale-like to 2 mm long. Male and female flowers separate; male flowers in clusters of 4–8 pairs, the female solitary or in pairs. Fruit globose, berry-like and red, the **seed completely enclosed**. Local in dry, sandy and rocky habitats.

1. *Araucaria heterophylla*
2. *Ephedra fragilis*

ANGIOSPERMS: **Seed plants bearing true flowers with ovules enclosed within an ovary or carpel.**

BASAL ANGIOSPERMS: Ancient group of monophyletic orders and families of angiosperms formerly classified as dicotyledons.

Order: PIPERALES

ARISTOLOCHIACEAE

Family of perennial herbs and woody climbers with creeping rhizomes. Leaves alternate, untoothed. Flower a tube with a limb at the apex; attractive to tiny insects that become trapped by downward-pointing hairs in the swollen chamber at the base of the tube. Inferior ovary, stamens 6, fused to the style in a column.

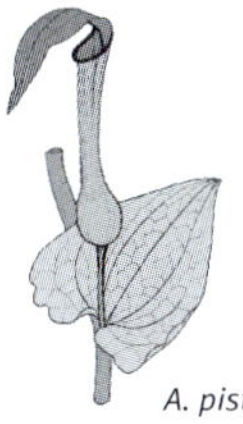

A. pistolochia

Aristolochia pistolochia A hairy perennial with a stock of numerous tubers. Leaves ovate-triangular, short-stalked, and with **minute teeth** on the margin and lower surface. Flowers almost **straight-tubed**, brownish with a **large, dark brown limb** that droops when mature. Dry, rocky and scrubby places; local.

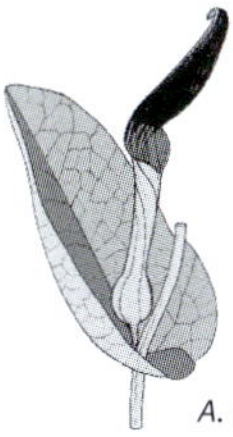

A. rotunda

Aristolochia rotunda Similar to *A. pistolochia*, but with ovate to heart-shaped **leaves that clasp the stem**. Flowers more or less straight-tubed, yellowish striped with brown, and with a large, dark brown limb that droops when mature. Dry, rocky and scrubby places; local.

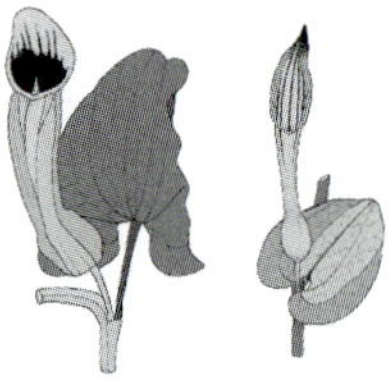

A. paucinervis A. pallida

Aristolochia paucinervis Leaves ovate-triangular, often with somewhat wavy edges, and with **petioles equalling the length of the pedicels**. Flowers erect and large relative to leaf-size, pale green-yellow or brownish with **contrasting dark purple stripes**. Local on sea cliffs; common on shale slopes near Castelejo. (***Aristolochia pallida*** occurs in the eastern Mediterranean and is superficially similar but with broader, less triangular leaves with petioles shorter than the pedicels.)

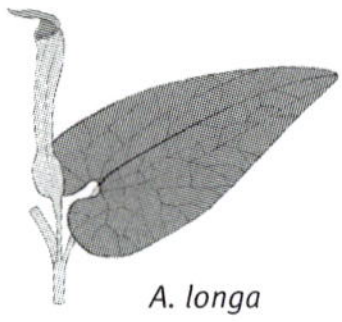

A. longa

Aristolochia longa BIRTHWORT A **more or less hairless** perennial with long scrambling or climbing stems. Leaves flat, ovate-triangular and with petioles equalling the length of the pedicels. Flowers rather inconspicuous, more or less straight-tubed, pale brownish-green with a **small, darker brown limb**. Populations traditionally described in the region may in fact be referable to the similar *A. paucinervis*, and appear to differ only in subtle leaf characteristics (leaves broad, flat without wavy margins), which may be a attributable to ecology. Damp, shady places or scrub. Locally common in woods near Monchique.

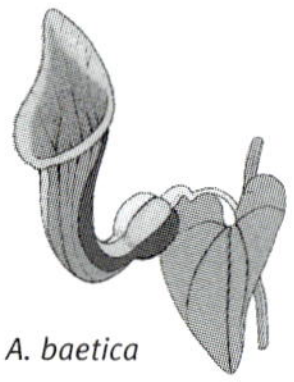

A. baetica

Aristolochia baetica Stems scrambling or climbing with **glaucous**, heart-shaped leaves with a deep cleft at the base. **Flowers rather large**, to 30 mm long, **maroon, brown or blackish-purple**, strongly curved and densely hairy within. Fruit large, to 50 mm long and hanging; green, later yellow then blackish and splitting to shed seeds. Damp and shady places, often near water, or among trees and scrub on *matos*. Locally common.

Order: MAGNOLIALES

MAGNOLIACEAE

Family of trees in which flowers are arranged in rings, with stamens and pistils in spirals on a conical receptacle, an arrangement seen in fossil flowers.

Magnolia grandiflora An exotic tree up to 20 m tall (usually less) with **simple**, ovate, leathery dark green leaves to 20 cm long, and large, white, scented flowers to 30 cm across. Planted in towns; a particularly large tree grows in Monchique.

Order: LAURALES

LAURACEAE

A family of mainly aromatic, evergreen shrubs and trees. Flowers small and cosexual. Fruit a berry.

Persea americana AVOCADO A tree to 15 m tall with large, ovate-elliptical leaves, leathery and pointed at the apex. Flowers small, white, with 5 petals. Fruit an avocado pear. Frequently planted in gardens in the region.

1. *Aristolochia paucinervis*
2. *Aristolochia paucinervis*
3. *Aristolochia paucinervis*
4. *Aristolochia longa*
5. *Aristolochia longa*
6. *Aristolochia baetica*
7. *Aristolochia baetica* fruits

MONOCOTS **[Monocotyledons]:** Second largest monophyletic group of angiosperms. Plants are typically characterised by having a single cotyledon (seed leaf), leaves with parallel (not net) veins, and flower parts typically in multiples of three. Mostly herbaceous and often with underground storage organs, such as bulbs and corms.

Order: ALISMATALES

ARACEAE

Perennial, usually hairless, tuberous perennial herbs with leaves all basal with distinct petioles. True flowers tiny, borne in a compact spike (spadix), enfolded in a large, often leafy bract (spathe). Fruit a berry.

A. italicum

Arum italicum **ITALIAN ARUM** A spring-flowering perennial with a large, **oblong** tuber and long-stalked leaves appearing in autumn. Leaves large to 35 cm long, and arrow-shaped with pointed lobes, held semi-erect on long stalks; mid to dark green with varying degrees of paler marbling and venation. Inflorescence typically arum-like; tiny **unisexual** flowers densely clustered around the base of the spadix enclosed by the spathe below; the visible portion of the spadix pale yellow; spathe large to 40 cm, later forward-drooping, and pale yellow-green. Berries green, ripening yellow then red, borne in a densely clustered spike. Common in woods at higher altitude, locally common in damp hedgerows and ditches elsewhere.

1. *Arum italicum*
2. *Arum italicum* fruits
3. *Arisarum vulgare*
4. *Zantedeschia aethiopica*
5. *Baldellia ranunculoides*

A. vulgare

Arisarum vulgare **FRIAR'S COWL** An autumn-flowering, short perennial with all basal leaves, broadly heart-shaped with rounded basal lobes. Flower stalk more or less equal to the leaf stalks and sometimes spotted. Spadix greenish or brown, and slightly protruding from the **hooded spathe**, which is white, striped green to chocolate brown; tubular-cylindrical below and forward-drooping above. Berries greenish to blackish. Very common on rocky ground and open *matos*, particularly near the coast.

B. tenuifolium

Biarum tenuifolium A small, autumn-flowering perennial to 20 cm with leaves appearing after flowering time. Leaves narrowly **oblong** and **linear** with wavy margins. Inflorescence sprouting direct from the ground and stalkless; **spadix often long-exceeding the spathe**, spadix long and slender, brown; spathe brownish or purplish. Low *matos* and rocky slopes; distribution in the region not fully known, and possibly under-recorded.

Zantedeschia aethiopica **ARUM LILY** A hairless, bushy perennial with rather fleshy, arrowhead-shaped leaves borne on long stalks. Flower stalks somewhat exceeding the leaves; spadix yellow and half as long as the spathe, the individual flowers indistinct; spathe pure white. Berries yellow, rarely produced. Widely planted and commonly naturalised in damp wood margins and along roadsides near Monchique.

ALISMATACEAE

Annual or perennial aquatic herbs with erect or floating leaves. Leaves stalked, with linear to rounded blades. Flowers cosexual and regular, arranged in panicles or racemes; stamens 3–6 or more; superior ovary. Fruit a head of achenes.

B. ranunculoides

Baldellia ranunculoides **LESSER WATER-PLANTAIN** An erect or spreading perennial to 20 cm tall with mostly basal, linear-**lanceolate** leaves, tapered at both ends and long-stalked. Flowers pale purplish-white, to 16 mm wide, solitary or in few-flowered clusters. Local in wet ditches and seasonally flooded plains.

JUNCAGINACEAE

A small family of aquatic herbs with linear, sheathing basal leaves. Flowers small and green, borne in erect spikes or racemes; flowers 3-parted, and carpels either 3 or 6, joined to a superior ovary. Fruit a capsule.

Triglochin bulbosa **BULBOUS ARROW-GRASS** A low to short perennial with a bulbous rootstock. Leaves few, linear, tapered to a point and less than 4 mm wide. Flowers tiny and greenish, to 2.5 mm, borne in slender spikes to 10 cm long. Fruit elliptical and spreading, to 12 mm long. Local in salt marshes and on river banks.

CYMODOCEACEAE

A small family, superficially similar to Zosteraceae, in which species are characterised by very narrow leaves and inconspicuous flowers with tiny petals (or petals absent); monoecious.

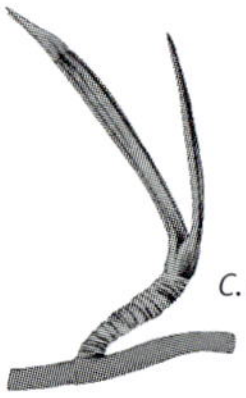
C. nodosa

Cymodocea nodosa Rather like an eelgrass; leaves 2–7 together on short shoots; stems with annual scars at the nodes from previous fallen leaves. Leaves dark green, to 40 cm long, and **narrow** to just 4 mm wide, 7–9-veined and spiny-**toothed** towards the tips. Flowers solitary, **without a perianth**; male flower with a filament-like stalk, **stamens 2**; female flower stalkless with **2 styles** and **2 thread-like stigma lobes**. Fruit laterally compressed. Local on submerged sandy substrates.

ZOSTERACEAE SEAGRASSES

A small family of perennial grass-like herbs (eelgrasses), unusual for being wholly aquatic in habit. Flowers reduced to a single stamen and style, borne on separate male and female flowers.

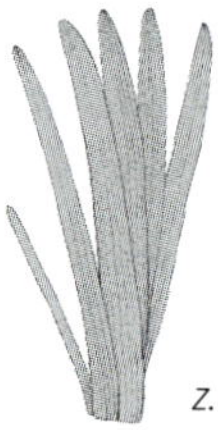
Z. marina

Zostera marina COMMON EELGRASS A submerged marine perennial with leaves to 50 cm long (sometimes longer) and **broad**, to 10 mm wide, with 3–9 veins, and broad, rounded, **bristle-tipped points**; basal sheaths **entire**, not split. Flowers greenish, borne in terminal, much-branched inflorescences, enclosed in the base of the sheath; stigma twice the length of the style; female flowers with **1 style**, male flowers with **1 stamen**. Locally abundant in suitable habitats; fine sandy or silt substrates, submerged to 10 m.

Order: DIOSCOREALES

DIOSCOREACEAE

A primarily tropical family including the yam. Dioecious: flowers small and greenish borne in spikes or racemes in the leaf-axils.

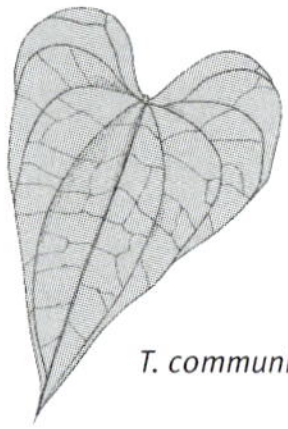
T. communis

Tamus communis BLACK BRYONY A tall, twining perennial climber to 4 m, dying back annually to a tuber; superficially similar to *Smilax*, but **spineless**. Leaves glossy, dark green, heart-shaped and long-stalked. Flowers greenish-yellow, to 6 mm, borne in loose racemes; male flowers with 6 stamens, female flowers with 6 minute lobes and a conspicuous ovary. Fruit a rounded red berry. Locally common in damp woods in the Monchique area.

Order: LILIALES

SMILACACEAE

Climbing lianas with heart-shaped leaves, hooked spines, and paired tendrils at the leaf bases.

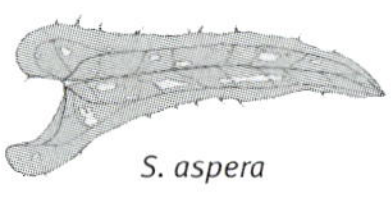
S. aspera

Smilax aspera COMMON SMILAX A dioecious, variable, creeping, scrambling or climbing shrub to 15 m with angled, smooth or prickly stems. Leaves dark shiny green and leathery, triangular to heart-shaped, often with prickles on the margins, and with a pair of tendrils at the base of the leaf stalk. Flowers green-white or yellowish or pinkish, and scented, to 5 mm, borne in branched clusters. Berry turning red through black when ripe. Very common on the *matos* and coastal scrub.

LILIACEAE

The traditional Liliaceae included a great number of genera now included in other families, including the Xanthorrhoeaceae, Colchicaceae, Asparagaceae, Amaryllidaceae and Ruscaceae. The newly described Liliaceae is much reduced. Species with a bulbous, tuberous or rhizomatous stock, and leaves with parallel veins. Perianth (calyx + corolla) usually 6-parted, often with tepals (petals and sepals all similar) separate or fused; stamens 6, style 1 or 3. Fruit a capsule or berry.

Tulipa **TULIPS** Bulbous perennials with a solitary stem, and few, rush-like leaves. Flowers conspicuous and typically tulip-like; tepals all alike; stamens 6. Fruit a 3-parted capsule containing fairly large, flat seeds. A genus better-represented in the Eastern Mediterranean.

T. sylvestris

Tulipa sylvestris **WILD TULIP** A short to medium, hairless, bulbous perennial to 45 cm tall. Leaves 2–3, strap-shaped and channelled. Flowers solitary, and nodding in bud; yellow tinged with orange and/or green, the tepals becoming recurved with age; stamens hairy at the base. Fruit a capsule, rarely produced. Generally very common at low altitude on the *matos*, rocky slopes, scrubs and roadsides. **Subsp. *australis*** (also treated as a distinct species) has slender stems, less than 2 mm in diameter, and flowers tinged crimson on the outside; fruits often produced.

1. *Zostera marina* PHOTO: TIM RICH
2. *Smilax aspera*
3. *Smilax aspera*
4. *Tamus communis*
5. *Tulipa sylvestris* subsp. *australis*

Fritillaria Small, bulbous perennials with solitary, unbranched stems. Leaves alternate, opposite or whorled, untoothed and mostly on the stem. Flowers often unpleasant-smelling, tubular or bell-shaped, nodding; tepals 6, all petal-like; stamens 6; style 3-lobed. Fruit an erect, 3-parted capsule containing many flat seeds.

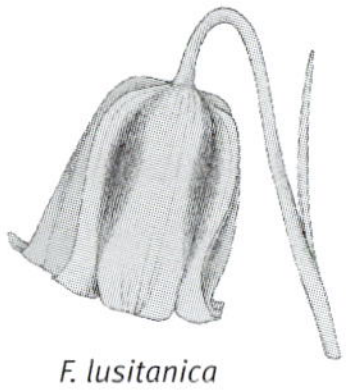

F. lusitanica

Fritillaria lusitanica A short perennial to 30 cm with erect stems and all leaves alternate. Leaves 5–9 and linear. Flowers 1–3; broadly bell-shaped, and large relative to the plant, to 40 mm long, green and brownish-purple, often slightly chequered. Fruit a capsule not winged. Local on low *matos*; fairly common on Cape St. Vincent.

COLCHICACEAE

A recently described family of some 200 species of herbaceous perennials with rhizomes or corms, previously included in the Liliaceae.

Colchicum AUTUMN CROCUSES Perennials with corms with basal clusters of leaves that appear either with, or after, the flowers. Flowers typically crocus-like with 6, spreading tepals; stamens 6; styles 4. Fruit a capsule, borne centrally in the leaf tuft. A genus better-represented in the Eastern Mediterranean.

C. lusitanicum

Colchicum lusitanicum A low perennial with flowers borne in the autumn, before the appearance of the leaves. Leaves 4–5, linear-lanceolate and dark, shiny green. Flowers pink and slightly **chequered with tepals to 60 mm long; anthers purplish or pink** (sometimes yellow), styles whitish and curled at the tips. Local on dry hillsides; inconspicuous except when in flower. ***Colchicum neapolitanum*** is very similar but with 2–4 leaves (not 4–5), and **smaller** pink flowers with tepals 30–45 mm long, **not or scarcely chequered with anthers always yellow**. Grassy and rocky hill slopes.

1

2

3

C. neapolitanum

Merendera filifolia A low cormous, autumn-flowering perennial, similar to *Colchicum* but with tepals free from one another and the anthers fixed at the base (not centrally). Flowers pink-mauve and crocus-like, solitary with anthers to 8 mm long. Local on dry, stony and bare ground.

Order: ASPARAGALES

ORCHIDACEAE ORCHIDS

One of the largest families of flowering plants, best represented in the tropics, with about 26,000 species worldwide. Perennials with tubers or rhizomes and flowers borne in spikes or racemes; perianth with 2 whorls of 3: the outer 3 all similar, the inner 3 with the central the largest and distinct, known as the labellum; ovary inferior; anthers and stigma together form a central column.

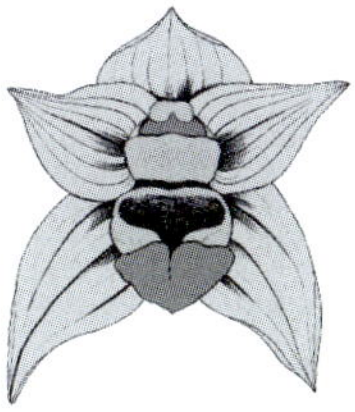

E. tremolsii subsp. *lusitanica*

Epipactis helleborine **BROAD-LEAVED HELLEBORINE** An erect, perennial orchid to 80 cm forming clumps with up to 3 stems. Lowermost leaves small, those around the centre of the stem largest and **ovate-elliptical** to 17 cm long, strongly veined and spirally arranged. Raceme many-flowered (with up to 100 flowers) and rather 1-sided; flowers opened wide but with petals and sepals incurved and held horizontal or slightly nodding; green, tinged purple. Long recorded in the area but probably confused with *E. tremolsii* subsp. *lusitanica*. ***Epipactis tremolsii* subsp. *lusitanica*** (= *E. lusitanica*) has widely **ovate-round** leaves with wavy edges. Flowers with sepals and petals **spreading to wide open, scarcely incurved**, the inner surface of the hypochile green. Common in pine woods in the region, possibly under-recorded if confused with the previous species.

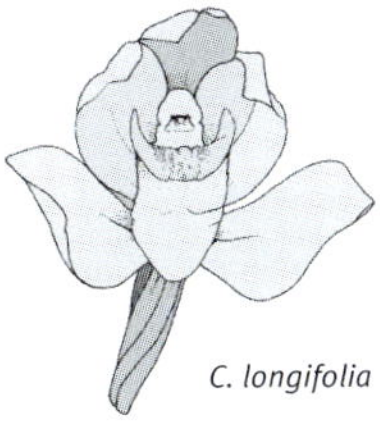

C. longifolia

Cephalanthera longifolia **SWORD-LEAVED HELLEBORINE** An erect, hairless perennial to 60 cm with stems with whitish scales below, and long, **narrow, linear leaves** around the stem, with the tips somewhat drooping. Flowers pure white, with orange markings inside the labellum, borne in dense spikes of 16–20; perianth segments pointed and half-opening. Local in damp, rocky woods in the Serra de Monchique.

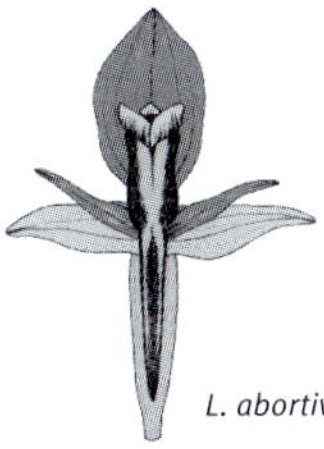

L. abortivum

Limodorum abortivum **VIOLET LIMODORE** A distinctive, purplish mycoheterotroph (plant lacking chlorophyll and parasitising a fungal symbiont) to 80 cm tall **without leaves**. Flower spike lax with up to 25 flowers; perianth segments long to 20 mm, violet or whitish; labellum yellowish or white with violet veins; bracts exceeding the ovaries. Local in cork oak forests on higher ground.

1. *Fritillaria lusitanica*
2. *Epipactis tremolsii* subsp. *lusitanica*
3. *Limodorum abortivum*

O. anthropophora

Orchis anthropophora (= Aceras anthropophorum) **MAN ORCHID** A short, slender orchid to 30 cm tall with 4–9 oval-shaped, blunt, shiny green leaves forming a basal rosette, and sheathing the stem. Inflorescence many-flowered, cylindrical and slender; flowers green-yellow streaked with dull red; sepals and petals forming a loose hood above the lip; lip to 15 mm long and pendant, with 2 slender 'arms' and 2 shorter, spreading 'legs' and with 2 swellings near the base; no **spur**. Local and uncommon; mainly central on rocky slopes and *matos*; rare in the west.

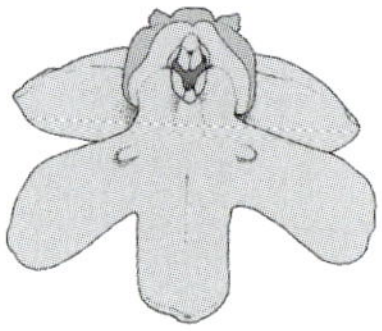

A. pyramidalis

Anacamptis pyramidalis **PYRAMIDAL ORCHID** An erect, robust perennial to 40 cm. Leaves lanceolate, grey-green and unspotted. Flowers bright pink, borne in distinctly **cone-shaped or dome-shaped dense spikes**; perianth all pink, the labellum broad and deeply 3-lobed, the spur long and slender. Readily distinguished by the shape of the inflorescence. Common on open scrub, particularly near the coast in the region.

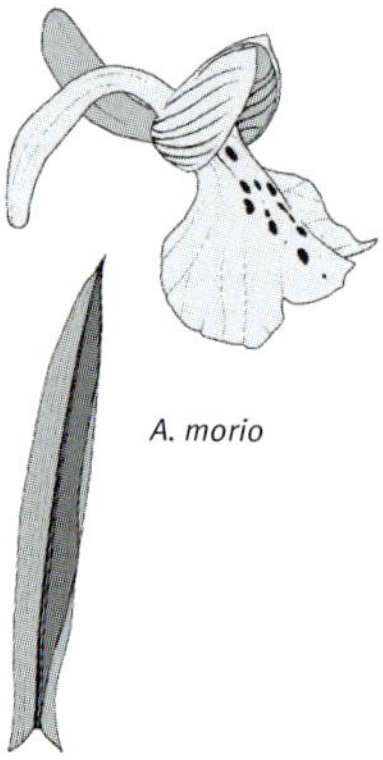

A. morio

Anacamptis morio (= Orchis morio) **GREEN-WINGED ORCHID** A low to medium perennial to 40 cm. Leaves mostly in a basal rosette, with several sheathing the stem and crowded, unspotted. Flowers usually pink (sometimes white), with perianth segments forming a hood, the lateral sepals strongly **veined and suffused in green**; labellum bluntly 3-lobed and almost folded in 2, with a central paler patch with red spots; spur equalling the ovary. Fairly common on scrub and roadsides, particularly on recently burnt sites; sometimes in large numbers. ***Anacamptis champagneuxii*** is very similar but has **lax inflorescences** with fewer flowers (typically 7–10) in which the labellum is longitudinally folded, flushed with pink-purple with a large **white, unmarked to scarcely marked** centre. Scattered in the region. ***Anacamptis longicornu*** has been described as distinct from the previous two species based on its more robust inflorescences with up to 20 flowers with a **long, more or less straight spur** to 17 mm. Scattered in the region.

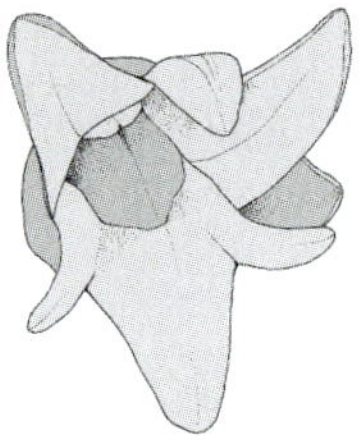

G. diphylla

Gennaria diphylla **TWO-LEAVED GENNARIA** A short, sometimes clumped perennial with 2 oval to heart-shaped leaves at the base, to 80 mm long. Inflorescence dense, 1-sided, to 13 cm long, with up to 40 flowers (usually fewer). Perianth segments all of similar length, to 4 mm with backwardly curved tips; greenish-yellow; lip 3-lobed, the middle lobe the largest; spur grooved. Local on coastal *matos*; common on Cape St. Vincent, local elsewhere.

S. aestivalis *S. spiralis*

Spiranthes aestivalis **SUMMER LADY'S TRESSES** A small perennial 10 cm tall with stems glandular above, with 3–6 basal, upright, linear-lanceolate leaves, flowering in early summer. **Inflorescence leafy below**, to 10 cm long with up to 20 flowers in a spiral; flowers pure white with perianth segments forming a tube that opens slightly at the end; labellum tongue-shaped and yellowish at the base. Uncommon in the region; on river banks and in other habitats near water. Recorded from the Monchique area. ***Spiranthes spiralis*** **(AUTUMN LADY'S TRESSES)** is similar, but taller to 30 cm, with leaves withered when in flower in early autumn, and the smaller flowers borne closely set in a **tight spiral on a leafless inflorescence**. Dry, shady, grassy and wooded habitats; local.

1. *Orchis anthropophora*
2. *Anacamptis pyramidalis*
3. *Anacamptis morio*
4. *Gennaria diphylla*
5. *Gennaria diphylla*
6. *Spiranthes spiralis* PHOTO: FINN RASMUSSEN

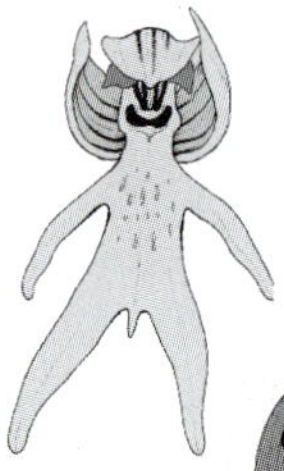

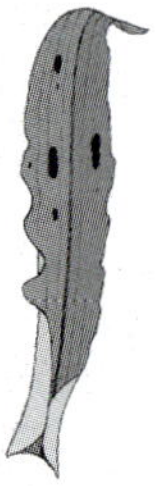

O. italica

Orchis italica **ITALIAN MAN ORCHID** An erect, robust perennial to 50 cm with 5–10 leaves in a rosette, and 2–4 sheathing the stem but not reaching the flowers. Leaves oblong-lanceolate, wavy-edged, and often flecked with brown. Flowers whitish-pink with darker veins, borne in dense, many-flowered inflorescences; perianth segments forming a loose hood; labellum to 20 mm long, tipped and spotted with purple, with **slender, pointed** 'arms' and 'legs' with a short 'tail' in the middle; spur down-curved and half the length of the ovary; bracts tiny, 1-veined and much exceeded by the ovary. Common on open scrub and pathways and in clearings in the *matos*, particularly near the coast. **_Orchis conica_** is superficially similar by virtue of its dense, cylindrical inflorescences with a 'whiskered' appearance. Flowers have pinkish-white perianth segments curving to form a hood, the sepals terminating in long, fine points and a 3-lobed labellum; flowers spotted with darker markings. Rather uncommon, sometimes in large numbers on limestone scrub.

O. olbiensis

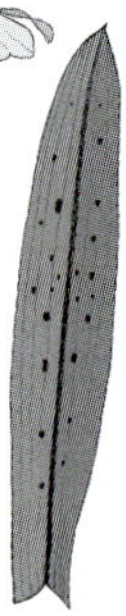

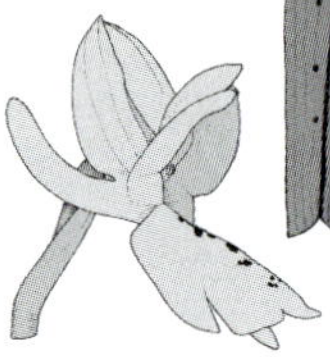

O. olbiensis

Orchis olbiensis **SOUTHERN EARLY PURPLE ORCHID** A medium-sized orchid to 40 cm tall. Leaves oblong-lanceolate, to 10 cm long, glossy green with **numerous longitudinal dark blotches**, sometimes unspotted. Flowers borne in loose, cylindrical spikes; perianth red-purple, the labellum shallowly 3-lobed with the lateral lobes deflexed, and with a paler central area spotted with red; **spur exceeding the ovary**, curved outwards, robust. Local in woods and scrub. Closely related to (and previously treated as a subspecies of) the North European species *O. mascula*, which differs in having denser inflorescences, and the spur equalling the ovary.

Orchis coriophora **BUG ORCHID** A rather small orchid to 40 cm tall. Leaves 4–10, lanceolate and folded. Inflorescence oblong and dense with many flowers. Flowers pinkish-brown and scented; perianth segments converging to form a **beaked** hood; lip spotted; spur downward-pointing and half the length of the ovary. Very rare; in the hills near Moncarapaho.

Orchis laxiflora **LOOSE-FLOWERED ORCHID** A rather tall orchid to 70 cm with up to 8 channelled, unspotted leaves. **Inflorescence lax**; stems red and flowers purplish-red with outwardly spreading sepals; petals incurved to form a loose hood; labellum 3-lobed, the central lobe shorter than the down-folded laterals, forming a tooth between them; centre of the labellum white and unspotted. Spur to two-thirds the length of the ovary and thickened at the end. Very rare; possibly only occurs just north of Portimão in grassland.

Orchis intacta **DENSE-FLOWERED ORCHID** A small orchid to 30 cm tall with normally heavily spotted leaves (sometimes unspotted) forming a rosette. **Inflorescence small and dense;** flowers small and scented; dull pinkish-white with purplish markings; sepals forming a hood; lip 3-lobed, the middle lobe rectangular and the laterals pointed. Rare in the Serra de Monchique in oak woodland.

1. *Orchis italica*
2. *Ophrys speculum*
3. *Ophrys speculum* subsp. *lusitanica*
4. *Ophrys speculum* subsp. *lusitanica*
5. *Ophrys speculum* × *Ophrys bombyliflora*
6. *Ophrys speculum* × *Ophrys bombyliflora*

Ophrys **BEE ORCHIDS** Distinctive tuberous perennials with insect-like flowers that dupe male insects of certain species into attempting to mate with the flowers to effect pollination (pseudocopulation). Sepals large and often greenish or pink; petals 2, smaller and hairy; labellum large and variously patterned, usually clothed in fur-like hairs. A complex genus with a centre of diversity in the Eastern Mediterranean; species in the Algarve appear to be visited only rarely by insects.

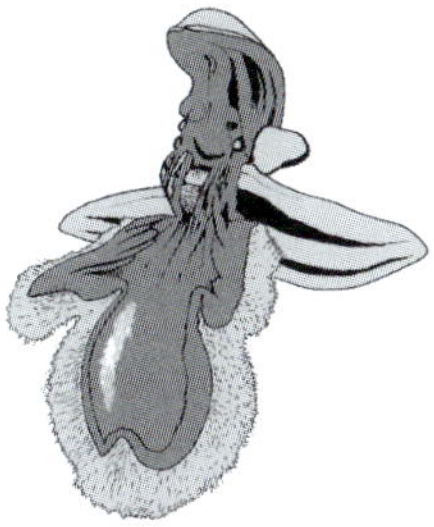

O. speculum

Ophrys speculum **MIRROR ORCHID** A short perennial orchid to 20 cm or less. Basal leaves oblong and blunt-tipped, the stem leaves pointed. Flowers borne in spikes of 1–3; sepals green or yellowish, usually striped brown; petals dark purple, hairy, and a third of the length of the sepals; labellum to 13 mm long, broad, 3-lobed, the middle lobe with a **blue, shiny speculum** framed with yellow, and fringed with brown or blackish hairs. Very common at low altitude in open *matos*, grassy habitats and stony pastures. **Subsp. *speculum*** has a **broad labellum without distinctly recurved margins**; **subsp. *lusitanica*** is larger, to 50 cm tall with up to 15 flowers, and the middle lobe of the labellum slightly elongated (oblong, not round) with recurved margins, the speculum with a deep yellow margin, and fringed with brownish red or yellowish (**not blackish**) hairs. Habitats similar to those of subsp. *speculum*; local and uncommon.

1

2

3

4

5

6

O. bombyliflora

Ophrys bombyliflora BUMBLEBEE ORCHID A short, loosely clump-forming perennial to 20 cm or less. Basal leaves ovate to lanceolate, forming a flat rosette, stem leaves erect and clasping the stem. Flowers borne in short spikes of 2–5; sepals green; petals green with a purplish base, triangular, and one third the length of the sepals; **labellum small**, to 8 mm long, 3-lobed with the lateral lobes deflexed, brown with a central bluish, shield-shaped speculum. Very common at low altitude in open *matos*, grassy habitats and stony pastures. Hybrids with *O. speculum* have been recorded in Salema.

O. lutea

Ophrys lutea YELLOW BEE ORCHID A short perennial orchid to 30 cm or less. Basal leaves oblong and pointed, stem leaves smaller and narrower. Flowers borne in spikes of 2–7; sepals green, petals greenish or yellowish and half the length of the sepals; labellum to 19 mm across, 3-lobed, with a broad, flat, **yellow margin** surrounding a red-brown area with a blue-grey speculum. Very common at low altitude in open *matos*, grassy habitats and stony pastures. **Subsp. *lutea*** has a labellum to 19 mm wide with a yellow margin 3–6 mm wide. The common form, **subsp. *galilaea*,** has a **smaller labellum** to 12 mm wide with a **narrow yellow margin** to 2 mm across. Very local in *Cistus*-dominated communities south of Paderne, possibly under-recorded.

O. tenthredinifera

Ophrys tenthredinifera SAWFLY ORCHID A short to medium perennial to 45 cm. Basal leaves ovate to lanceolate and blunt or pointed, stem leaves similar. Flowers borne in short spikes of 3–8; **sepals purplish-pink** (rarely green or white), the central erect; petals similar in colour and 1/3 the length of the sepals; **labellum broad and square, scarcely lobed or unlobed** to 14 mm, brownish-purple with a broad yellowish margin, and a small, brown-spotted, 2-parted speculum. Locally frequent in dunes and low *matos*, mostly coastal.

O. fusca

Ophrys fusca SOMBRE BEE ORCHID A short to medium perennial to 40 cm. Basal leaves oblong-lanceolate and blunt-pointed, stem leaves smaller and narrower. Flowers borne in spikes of 3–10; sepals green or pinkish; petals green and half the length of the sepals; labellum to 15 mm, horizontal to down-curved, 3-lobed and purplish- or yellowish-brown, often yellow-edged, and with a bluish or greyish W-shaped speculum. Frequent on coastal cliffs in the Cape St. Vincent area and on coastal *matos* elsewhere. This species is very variable and has been the subject of much investigation. Several distinct variants occur in the Algarve but their specific rank remains equivocal. **Subsp. *iricolor*** can be distinguished by the lip, which is wine-red underneath, and a bluish, shining speculum; the lip of **subsp. *fusca*** is pale green underneath. Local populations in which flowers have a very distinct W-shaped band may be referable to the morphologically similar ***O. omegaifera* subsp. *dyris*.**

O. apifera

Ophrys apifera BEE ORCHID A short to medium perennial to 50 cm. Basal leaves ovate to lanceolate and blunt or pointed, stem leaves similar but smaller. Flowers borne in short spikes of 2–10; sepals bright pink with a green mid vein (rarely green or white); petals green or purplish and less than 1/3 the length of the sepals; labellum 3-lobed, the **central lobe curved backwards**, with a shield-shaped brown or violet speculum with a yellowish margin. Local and sporadic on sea cliffs, and disturbed environments such as road cuttings inland.

1. *Ophrys bombyliflora*
2. *Ophrys lutea*
3. *Ophrys lutea*
4. *Ophrys tenthredinifera*
5. *Ophrys tenthredinifera*
6. *Ophrys fusca*
7. *Ophrys fusca*
8. *Ophrys apifera*

O. scolopax

Ophrys scolopax **WOODCOCK ORCHID** A short to medium perennial to 45 cm. Basal leaves lanceolate and pointed, stem leaves narrower and more pointed. Flowers borne in rather long spikes of 3–12; **sepals pink** (rarely green or white), the central erect; petals similar in colour and half the length of the sepals; labellum to 12 mm, ovate, 3-lobed, and brownish-purple and velvety with a large brownish-blue, H-shaped or spotted speculum with a narrow yellow margin. Locally frequent in open habitats at low altitude. **Subsp. *scolopax*** has triangular sepals, and the **middle lobe of the labellum not markedly narrowed** above the base; **subsp. *apiformis*** has more or less **linear sepals**, and the **labellum narrowed** (strongly recurved) above the base.

Serapias Similar to *Ophrys* species but with a larger labellum that is downward-pointing and tongue-like, with 2 short, upturned lateral lobes. Potentially hybridising with other genera including Ophrys, the extent of which is unknown in the region.

S. lingua

Serapias lingua **TONGUE ORCHID** A short perennial to 25 cm tall. Leaves narrowly lanceolate and grey-green. Flowers borne in lax spikes of up to 9; sepals and petals purple; labellum to 32 mm long, maroon-coloured with a single blackish protuberance at the base, the prominent central lobe rather narrow and pointed. Common and widespread on the *matos* and in damp or flood-prone grassy habitats. Some populations in the region may in fact be referable to ***Serapias strictiflora***, which is similar but has a **narrower labellum with straight margins**; further investigation is required. ***Serapias parviflora*** is similar to *S. lingua* but with smaller flowers; the labellum with **2 (not 1)** brownish-red (not blackish) ridges at the base. Limestone areas, damp grassland and marshes; local.

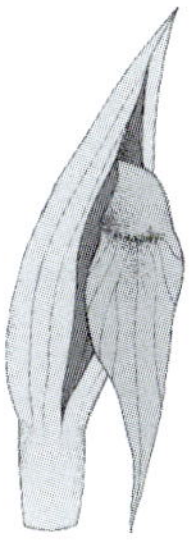

S. parviflora

Serapias cordigera **HEART-FLOWERED ORCHID** A short to medium perennial to 50 cm tall. Leaves narrowly lanceolate, channelled, and sharply pointed; the lowermost leaves sheath-like, purplish and spotted at the base. Flowers borne in short spikes of up to 10; sepals and petals reddish; labellum to 35 mm long, reddish, the prominent central lobe **broad and heart-shaped**, and twice as large as the sepals, hairy at the mouth, and with 2 dark ridges at the base. Damp grassland and marshes; local.

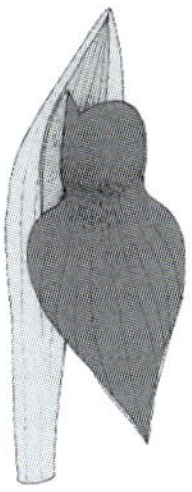

S. cordigera

Serapias vomeracea **LONG-LIPPED SERAPIAS** A short to medium perennial to 50 cm tall. Leaves linear, channelled, grey-green, and sharply pointed; the lowermost leaves sheath-like, and green (not purple). Flowers borne in short spikes of up to 10 with equal or longer bracts; sepals and petals purplish; labellum to 40 mm long and much exceeding the sepals; usually **pale** yellowish-red to maroon, with 2 parallel, similar-coloured ridges at the base; the prominent middle lobe triangular. Damp grassland and marshes; local.

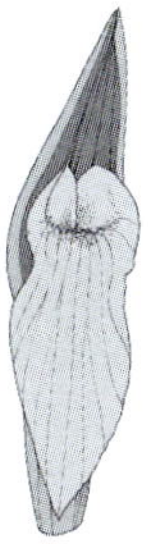

S. vomeracea

1. *Ophrys scolopax*
2. *Serapias lingua*
3. *Serapias lingua* pollinator
4. *Serapias cordigera*
5. *Serapias cordigera*
6. *Serapias vomeracea*

IRIDACEAE

Bulbous, tuberous and rhizomatous perennials, often with linear leaves, all basal or alternate. Flowers often distinctly 3-parted, enclosed in 1 or 2, often papery spathes when in bud; tepals 6 and petal-like, all similar or 3 different; ovary inferior. Fruit a 3-parted capsule.

Iris Flowers with the outer 3 tepals (falls) horizontal to down-turned, and the inner 3 (standards) erect; styles rather petal-like and arching over the falls, each with a stamen beneath.

I. xiphium

Iris xiphium SPANISH IRIS A medium bulbous perennial to 60 cm with flowering stems exceeding the narrow, channelled leaves to 5 mm wide that appear in autumn. Flowers large and solitary, **violet-blue** with a yellow ring on the centre of the ovate falls (rarely all yellow-flowered variants, which have been described as *I. lusitanica* but which are doubtfully distinct at the specific level); standards erect and somewhat smaller than the falls. Fairly frequent in the region in rocky, grassy areas and scrub.

Iris planifolia A low to short, bulbous perennial usually under 20 cm tall. Leaves all basal, shiny green, produced in a fan-like arrangement. Flowers bluish-violet with darker veins, and **small standards,** to 20 mm long, and toothed. Local on limestone rocks and fields.

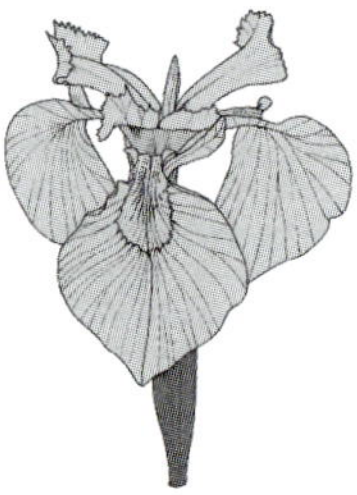

I. pseudacorus

Iris pseudacorus YELLOW FLAG A tall and erect rhizomatous perennial herb to 1.2 m. Leaves sword-shaped, to 90 cm long and 30 mm wide, green or greyish. Flowers large to 10 cm across, and **bright yellow** with faint red-purple veins, borne in clusters of up to 3, each with a papery spathe below; unbearded, and with standards markedly smaller than the falls. Capsule large and splitting into 3 segments when ripe, exposing large brown seeds. Local near seasonally running water.

I. subbiflora

Iris albicans A short, erect rhizomatous perennial herb to less than 60 cm tall with rhizomes thick and spreading partially above ground. Leaves broad and sword-shaped, to 45 mm wide. Flowers borne in clusters along the stems, each with a spathe that is green at flowering time, or papery at the tip; **flowers large and pure white,** the falls with yellow beards. Native to the Middle East; locally commonly naturalised and patch-forming on roadsides in the region. ***Iris subbiflora*** is fairly similar but with **violet-purple flowers,** the falls with a **white or violet beard**. Gardens and dry, open scrub.

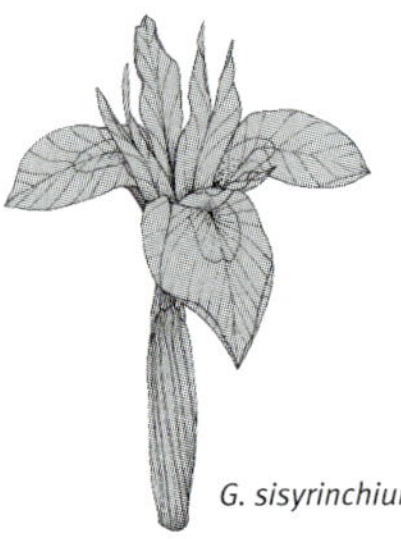

G. sisyrinchium

Gynandriris sisyrinchium BARBARY NUT Rather similar to *Iris* spp. but with a corm with a fibrous tunic; a short perennial to 30 cm with slender stems. Leaves 2, sheathing at the base, the free portion longer than the flowering stems, to 40 cm (usually less), linear and channelled. Flowers borne in groups of up to 5, violet-blue, the falls with white and yellow markings, and the standards slightly shorter. Common on pathways, roadsides and other dry, open habitats; often patch-forming. Flowers open in the afternoon sun.

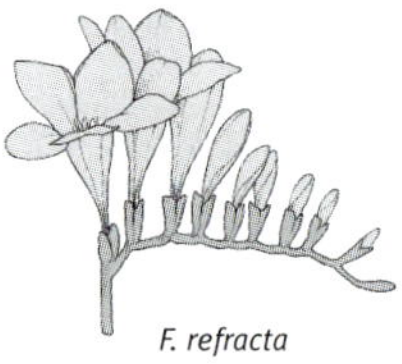

F. refracta

Freesia refracta A perennial with an underground corm and leafy, branched stems. Leaves 5–6, linear-lanceolate. Flowers yellow (rarely white), appearing tubular, and 2-lipped, fragrant. Native to South Africa and cultivated as an ornamental.

Crocus Small perennials arising from corms. Leaves basal, enclosed beneath by sheaths with a white strip running along the centre. Flowers cup-shaped, with 1–2 spathes; tepals 6, and all similar; stamens 3; style solitary with 3 many-branched stigmas. Fruit a capsule. Poorly represented in the region.

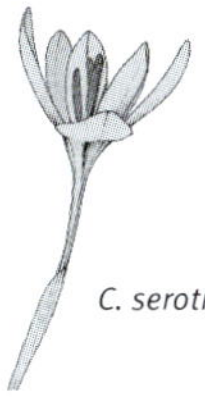

C. serotinus

Crocus serotinus A low cormous, autumn-flowering perennial with leaves partially developed when in flower or appearing immediately after. Leaves 4–7, narrow to 3.5 mm wide. Flowers 1–2, pale lilac with a pale yellow throat and yellow to orange stamens and style; scented. Usually on coastal sands.

R. ramiflora

R. columnae

Romulea bulbocodium A low, cormous spring-flowering perennial. Leaves 3–7, deep green and curved or straight and rush-like. Spathe mostly papery and weakly veined; flowers crocus-like, borne on short stems to 30 mm long, white to lilac with a yellow throat, and striped purple; tepals elliptical and pointed, to 35 mm long; stamens about half the length of the tepals, stigmas variable but usually **overtopping the anthers**. Common on coastal sands. **Romulea ramiflora** is very similar to *R. bulbocodium* but with spathes green (papery only at the margin) and closely and conspicuously veined, and the stigmas more or less equalling, or scarcely overtopping, the anthers. Damp, grassy habitats, usually coastal; uncommon. **Romulea columnae** has flowers off-white to pale violet with **stigmas below the top of the anthers**. Spathe papery and rust-spotted.

1. *Iris xiphium* PHOTO: TIM RICH
2. *Iris pseudacorus*
3. *Iris albicans*
4. *Gynandriris sisyrinchium*
5. *Romulea bulbocodium*
6. *Romulea bulbocodium*

Gladiolus Cormous perennials with fans of flat, sword-shaped and ribbed leaves. Flowers borne in long, rigidly erect stems, each flower with a green bract; tepals 6, unequal, fused into a short tube at the base; stamens 3. Fruit a capsule.

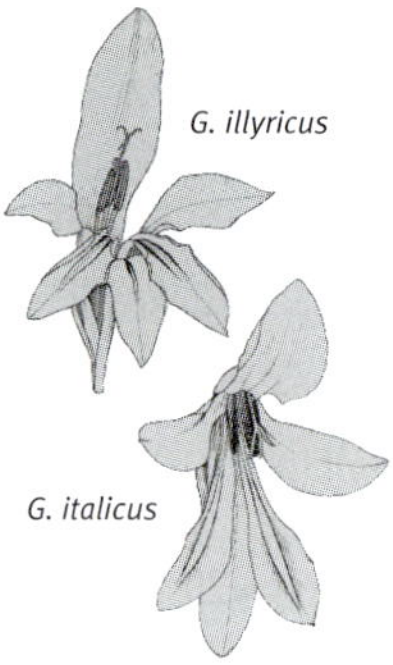

Gladiolus illyricus A tall, delicate, cormous perennial with sword-shaped leaves to 40 cm long. Flowers borne in a lax, weakly 2-sided inflorescence with 3–10 flowers, each flower with 2 bracts; flowers to 40 mm long, bright pink, the upper segments unequal, the middle longer than the laterals. Anthers **equalling or shorter than the filaments**. Seeds winged. Common on the *matos* and low scrub. **Gladiolus italicus** is very similar but with **anthers longer than the filaments**, or aborted, and **seeds not winged**. Cultivated ground; local and generally less common in the region than the previous species.

1. *Gladiolus illyricus*
2. *Asphodelus fistulosus*
3. *Asphodelus fistulosus*
4. *Asphodelus ramosus*
5. *Asphodelus ramosus*

XANTHORRHOEACEAE

A recently described family of species that were formerly included in the Liliaceae, with members characterised by a dense tuft of long narrow leaves, and a stout woody spike of flowers.

Aloe Succulent shrubs with robust, spiny-margined leaves forming a rosette. Flowers in racemes with tubular corollas; often orange or red. Fruit a capsule. Native to Africa and Madagascar but widely cultivated in the region.

A. mitriformis

Aloe arborescens TREE ALOE A **large, much-branched shrub** to 3 m. Leaves succulent and narrowly triangular-lanceolate, to 60 mm long, crowded in large rosettes and deflexed-spreading; greyish without pale markings, and toothed along the margin. Inflorescence unbranched and dense; flowers scarlet, to 40 mm long. Frequently cultivated, rarely far from habitation. ***Aloe mitriformis*** is similar but ground-hugging, with erect, branched inflorescences of red-orange tubular flowers borne in dense, rounded clusters. Planted in the region. ***Aloe vera*** has bluish or reddish leaves to 60 cm long, and **yellow flowers**, and is naturalised on maritime rocks and sands near habitation in the region.

ASPHODELACEAE

A family of perennial tuberous herbs, usually with a basal aggregation of stalkless, fleshy leaves. Flowers cosexual, actinomorphic or zygomorphic in multiples of three, and borne in spikes or panicles. Fruit a dehiscent capsule.

Asphodelus Robust, hairless, herbaceous perennials with linear leaves forming a basal tuft, and tall, leafless stems of star-shaped flowers, each with a papery bract. Fruit a 3-parted capsule. Unpalatable to livestock, and therefore a prominent feature in heavily grazed areas.

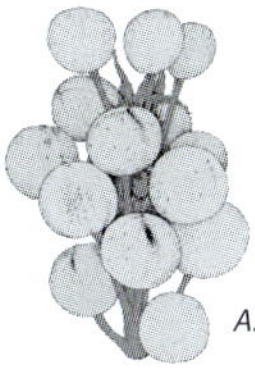

A. ramosus

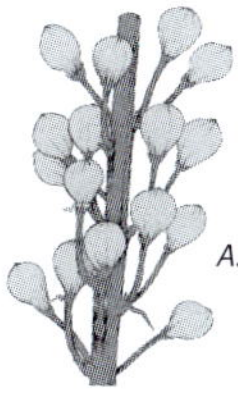

A. aestivus

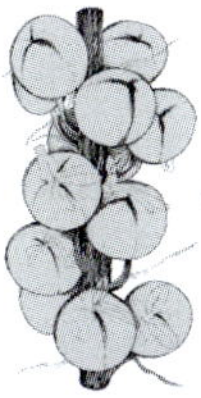

A. fistulosus

Asphodelus ramosus A tall, robust herbaceous perennial to 1 m with numerous fleshy roots. Leaves flat, strap-shaped and grey-green. Flowers white and star-like, to 30 mm across, the tepals with brownish mid-veins, borne in tall, erect, much-branched inflorescences in which the **lateral branches are almost as long as the terminal**. Capsules ovoid, **large** (8–14 mm). The most common species of *Asphodelus* in many parts of the region; often abundant in stony pastures and goat-grazed *matos*. ***Asphodelus aestivus*** is very similar and often confused with the above species but with **smaller, egg-shaped** (not ovoid) capsules to 7 mm long or less. Inflorescence branch length variable, and not a reliable diagnostic in the region. Similar habitats, and locally frequent.

Asphodelus fistulosus HOLLOW-LEAVED ASPHODEL A small, tufted **annual** or short-lived perennial to 60 cm with numerous fleshy roots. Leaves linear and **slender**, to 4 mm wide, **hollow**, and cylindrical. Flowers white or pinkish, borne in a lax raceme of 10–15 on a long, smooth stalk with membranous, whitish bracts; tepals with brownish mid-veins. Capsule to 5 mm in diameter. Common in grazed pastures, stony ground and waste places across the region.

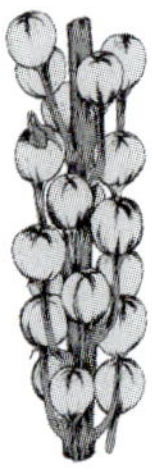

A. albus

Asphodelus albus **WHITE ASPHODEL** A medium to tall perennial to 1 m, superficially similar to *A. ramosus*. Leaves linear and grey-green, with a central **keel**. Inflorescence a *narrow*, **spike-like raceme**, rarely branched at the base, and usually **unbranched** above or with few, short lateral branches; flowers white, to 30 mm across, the tepals with greenish mid-veins. Capsules to 20 mm long, borne densely along the spike. Locally common on rocky ground and grazed *matos*.

AMARYLLIDACEAE

Bulbous perennials with leafless stems, similar to the Liliaceae. Leaves basal. Flowers solitary or in umbels, enclosed in papery bracts before flowering; flower parts in threes; tepals 6, usually all petal-like; ovary inferior; stamens 6. Fruit a 3-parted capsule.

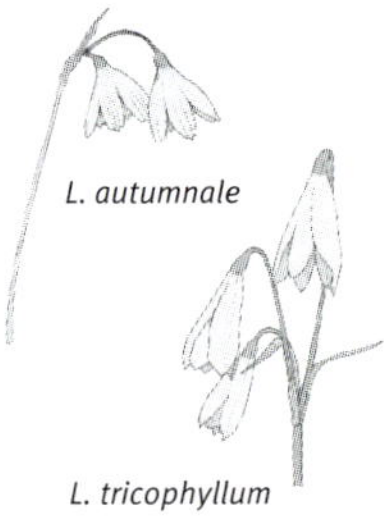

L. autumnale

L. tricophyllum

Leucojum trichophyllum **THREE-LEAVED SNOWFLAKE** A small, slender, spring-flowering bulbous perennial. Leaves usually 3, linear and narrow. Flowers white, to 20 mm long and bell-shaped, borne in clusters of 2–4 on slender stalks to 60 mm long on scapes to 30 cm; style thread-like and slightly longer than the stamens; spathe with 2 valves. Local in sandy habitats, usually coastal and varying considerably in abundance with seasonal rain. ***Leucojum autumnale*** is similar but **flowers in late summer**, with the thread-like **leaves appearing after (not during) flowering**. Rocky slopes and hillsides.

Narcissus Bulbous perennials with basal leaves and hollow, leafless flower stalks (scapes). Flowers with 6 tepals; stamens 6, surrounded by a cup- or trumpet-like corona. Capsule 3-parted.

N. papyraceus

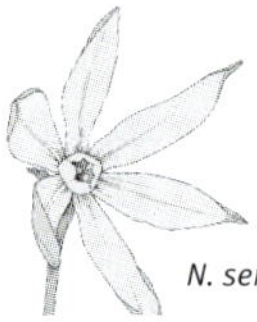

N. serotinus

Narcissus papyraceus **PAPERWHITE NARCISSUS** An autumn-flowering, hairless, bulbous perennial with broad, strap-like leaves, blue-green and slightly channelled, to 15 mm wide. Flowers pure white, borne in umbels of up to 20; flowers to 40 mm and strongly fragrant. Dry rocky places and fallow ground; local. ***Narcissus serotinus*** has very **narrow, cylindrical leaves** to 1 mm wide appearing in spring, and **absent when in flower** in the autumn; flowers white with a very small, orange corona. Dry habitats. ***Narcissus tazetta*** is similar but with flat, broad, **strap-like leaves present when in flower**, and flowers white, with a larger yellow corona that is twice as wide as high, borne on a 2-edged, flattened scape.

N. tazetta

N. willkommii

Narcissus bulbocodium A spring-flowering, short, tufted perennial. Leaves 2–4, more or less upright and dark green. Flowers held horizontally and with a **large, broadly cone-shaped corona**, to 25 mm long and bright yellow; perianth segments pointing forwards, and pointed; stamens inserted, not exceeding the corona; flowers borne on stalks slightly exceeding the leaves. Common in rocky places, coastal scrub and other dry open habitats across the region. ***Narcissus gaditanus*** is rather similar to *N. bulbocodium* but with leaves exceeding the scape, **small flowers** to 20 mm in diameter with a cup-like corona less than 9 mm wide, and **large, backward-pointing perianth segments** to 9 mm long; **flower stalks shorter than their spathes** when mature. Much decreased in recent decades; local on coastal cliffs. ***Narcissus willkommii*** is very similar to *N. gaditanus* but with larger flowers to 24 mm across, and **flower stalks exceeding their spathes** when mature (at least the longest). Local in damp grassland.

Pancratium maritimum **SEA DAFFODIL** A distinctive, late-summer flowering short to medium, clump-forming, bulbous perennial. Leaves fleshy, grey-green, hairless and long to 50 cm. Flowers large and white, to 15 cm long, fragrant, borne in umbels of 3–15; tepals all similar; stamens 6, borne on the rim of the cone-like corona. Fruit a 3-parted capsule. Common along the coast, and a distinctive component of the flora on maritime dunes.

1

2

3

4

5

6

1. *Leucojum trichophyllum*
2. *Leucojum autumnale* PHOTO: FINN RASMUSSEN
3. *Narcissus gaditanus*
4. *Narcissus bulbocodium*
5. *Narcissus bulbocodium* (marsh form)
6. *Pancratium maritimum* PHOTO: ANDY BAILEY

ALLIACEAE

A family of perennial bulbous herbs, with basal, stalkless leaves; species in the Algarve are garlic-scented when crushed. Inflorescence a cymose umbel of stalked, cosexual flowers, enclosed in a calyptra when in bud. Fruit a dehiscent capsule.

Allium Distinctive bulbous perennials smelling of onion or garlic. Flowers borne in terminal umbels that are often spherical; tepals 6 and all similar; stamens 6. Fruit a 3-parted capsule.

A. ampeloprasum

Allium ampeloprasum WILD LEEK A robust, leek-like bulbous perennial with stout flowering stems to 1 m tall. Leaves pale blue-green, V-shaped in cross-section, at least below, rough along the margin and with a central keel beneath; leaves often withered when in flower. Flower-heads very large and spherical with numerous lilac flowers with yellow anthers, borne on long stalks. Locally common on waste and fallow land, particularly near the coast. ***Allium porrom*** is the cultivated leek, which is believed to be derived from this species.

1. *Allium ampeloprasum*
2. *Allium ampeloprasum*
3. *Allium roseum*
4. *Allium subvillosum*
5. *Asparagus albus*
6. *Asparagus acutifolius*

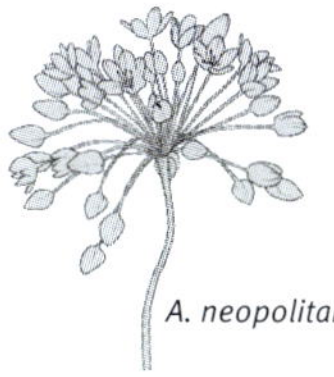
A. neopolitanum

Allium neapolitanum **Naples garlic** A short to medium bulbous perennial to 45 cm tall. Leaves few (2–3), linear, about 20 mm wide, sheathing the stem at the base and keeled along the back. **Flower stems triangular in cross-section** and smooth, carrying **white flowers** in loose clusters to 70 mm across; spathe 1-valved and shorter than the flower stalks. Locally common in grassy habitats and pastures. ***Allium roseum*** is similar to *A. neapolitanum* but with **cylindrical (not angled) stems** and pink flowers borne in loose umbels that are sometimes mixed with **bulbils**. Habitats similar to those of *A. neapolitanum*. ***Allium trifoliatum*** is similar to *A. neapolitanum* but with **leaves and sheaths hairy**, at least along the margins, cylindrical stems, and white flowers with pink mid veins along the tepals, or flushed pink throughout. Rocky habitats. ***Allium massaessylum*** is distinguished by its narrower leaves to 10 mm wide, more or less cylindrical stems, **2-valved spathe** that exceeds the flower stalks, and white flowers, the **tepal segments with a purple keel**.

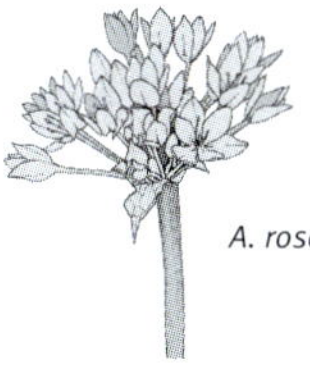
A. roseum

A. trifoliatum

Allium pallens A small, bulbous perennial to 40 cm tall. Leaves 3–4, to 20 cm long, linear and **very narrow to 2 mm wide**, sheathing the stem at the base. Flowers borne in compact, hemispherical umbels to 35 mm across; perianth to 6 mm, white or pale pink; stamens included within the perianth or with the yellow anthers slightly protruding; **spathe 2-valved**, persistent, and exceeding the flower stalks. Dry, open habitats.

A. pallens

Allium subvillosum A small bulbous perennial. Leaves 2–5, to 20 cm long, straight, narrowly linear and almost basal. Stem cylindrical. Flowers borne in hemispherical, many-flowered umbels; **perianth completely white** with yellow anthers inserted or slightly exserted; spathe 1-valved and persistent. Common on maritime sands.

ASPARAGACEAE

A recently described family comprising species formerly included within the Liliaceae.

Asparagus Shrubby, hairless, rhizomatous perennials with tough stems, becoming woody. Leaves replaced by cladodes (reduced stems). Flowers small and inconspicuous, bell-shaped with 6 tepals. Fruit a small berry.

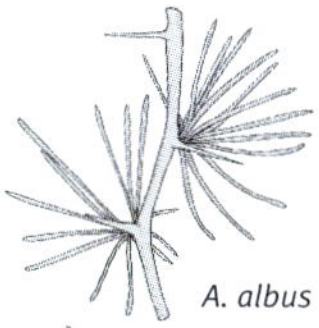
A. albus

Asparagus albus A tall, autumn-flowering perennial with woody, whitish, flexible stems, smooth or grooved; **cladodes not leaf-like, long (to 25 mm)**, soon-falling to leave bare, spiny stems. Flowers small, to 3 mm, and sweetly scented, borne in clusters of 6–15 on stalks to 5 mm long; **perianth white** with 6 spreading lobes. Fruit a berry turning red through black. Frequent on the *matos*.

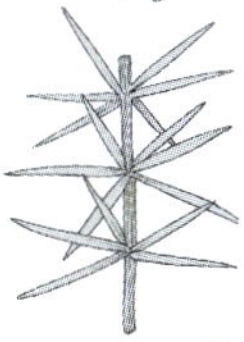
A. acutifolius

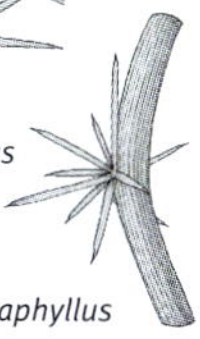
A. aphyllus

Asparagus acutifolius Similar to *A. albus*; a **climbing subshrub** with much-branched, woody stems to 2 m. **Cladodes more or less equal** in clusters of 10–30, **small to just 8 mm long**, and **spine-tipped. Flowers borne in clusters of 2–4**, to 4 mm long, **pale yellowish-green**, sweetly scented and mixed among the cladodes. Berry black when ripe. Frequent on the *matos*. ***Asparagus aphyllus*** is similar to *A. acutifolius* but **shorter and non-climbing**, to less than 1 m tall with hairless green stems. **Cladodes**

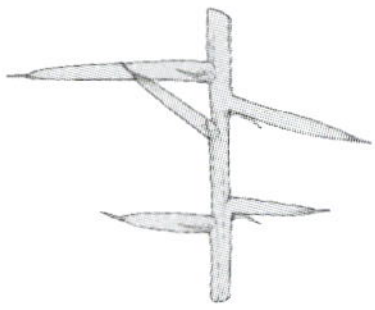

A. stipularis

markedly unequal, in clusters of 3–7, to 20 mm long. Flowers in groups of 3–6. Berries black when ripe. Rocky scrub and on the *matos*. ***Asparagus stipularis*** is similar to *A. acutifolius* but with very spiny **cladodes which are solitary** or in clusters of 2–3.

Cordyline australis A small tree up to 10 m (usually less) with a stout grey trunk that is simple or branched. Leaves sword-shaped, erect to 1 m long. Flowers white, scented and borne in dense panicles to 1 m long. Native to New Zealand but widely planted in towns and gardens in the Algarve.

Ornithogalum **STARS OF BETHLEHEM** Bulbous perennials with basal leaves in a rosette. Flowers normally white and star-shaped, borne in racemes; tepals separate and petal-like; stamens 6. Fruit a 3-valved capsule with many seeds.

O. narbonense

Ornithogalum narbonense A medium, spring-flowering perennial. Leaves 4–6, linear with a sheathing base, to 16 mm wide, and persistent until after flowering. Flowers white and scentless, borne erect on stalks the same length in a **tall, many-flowered, slender raceme**; perianth segments wide and spreading, to 16 mm long with a prominent green strip along the back; anthers yellow; ovary to 5 mm with a flattened top, **distinctly exceeded by the narrow style**. Common in rocky pastures.

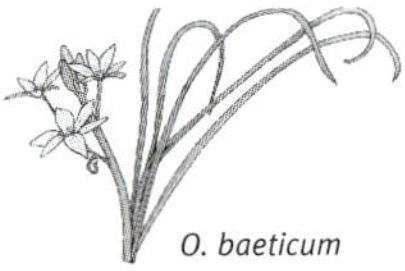

O. baeticum

Ornithogalum baeticum A medium, spring-flowering perennial. Leaves 6–9, linear and flat or slightly keeled, **exceeding the inflorescence**. Flowers white and scentless, borne erect on **ascending** stalks with shorter or longer bracts; perianth segments with a green vein running separately that **terminates at the flower stalk** (not running down the flower stalk). **Capsule 6-angled and furrowed.**

O. arabicum

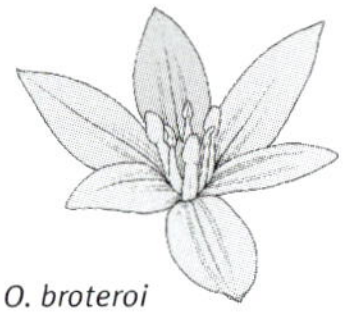

O. broteroi

Ornithogalum arabicum A stout, bulbous perennial to 70 cm with numerous offsets from the main bulb. Leaves 7–8, flat or channelled, plain green and sheathing at the base. Flowers white or cream, to 50 mm long with a **conspicuous violet-black ovary** in the centre, borne in a lax raceme with up to 25 flowers; flowers with **broad tepals without a central green stripe**. Capsule cylindrical. Local in dry, rocky places. ***Ornithogalum broteroi*** has a **single narrow leaf** to 6 mm wide, and few, white flowers without a green stripe on the back of the perianth segments, with a yellowish (not blackish) ovary. Local in dry, rocky places. ***Ornithogalum concinnum*** is similar but with **2–3 leaves** (not 1), which are **broader to 10 mm wide**.

Urginea Similar to *Ornithogalum* but with very large bulbs and numerous flowers borne in a long, terminal spike.

U. maritima

Urginea maritima **SEA SQUILL** A stout perennial to 1.5 m tall with a very large bulb to 15 cm across, sitting close to the surface. Leaves all basal, large and broadly lanceolate, slightly wavy-margined and shiny green, prominent in spring, but disappearing before the flowers appear in late summer. Flowers white and star-like, to 16 mm across, borne in a long, stout, spike-like inflorescence arising erect from dry ground. Common in a range of dry habitats, including coastal scrub and *matos*; a conspicuous plant in the region in later summer.

Scilla **SQUILLS** Bulbous perennials (squills) with basal leaves and leafless stalks. Flowers star-shaped and usually blue, borne in racemes or solitary, with a bract at the base of each flower stalk; tepals 6, separate and spreading. Fruit a 3-parted capsule.

S. autumnalis

S. monophyllos

Scilla peruviana A robust, medium, spring-flowering perennial to 50 cm. Leaves numerous, long and flat, to 35 mm wide and 60 cm long. Inflorescence a showy, hemi spherical cluster of up to 100 blue flowers; flowers blue and star-like with yellow anthers; bracts papery and long. Locally common, often in large numbers in damp, grassy habitats.

Scilla autumnalis **AUTUMN SQUILL** A low to short **autumn-flowering perennial** with stems to 25 cm, and few, hairless, narrowly linear leaves to only 3 mm wide, absent when in flower. Flowers up to 35, blue and star-like, borne in a spike-like raceme; bracts absent. Locally common on dry, rocky slopes.

1. *Cordyline australis*
2. *Ornithogalum narbonense*
3. *Ornithogalum broteroi*
4. *Urginea maritima*
5. *Scilla peruviana*
6. *Scilla autumnalis* PHOTO: FINN RASMUSSEN

1. *Scilla monophyllos*
2. *Bellevalia hackelii*
3. *Bellevalia hackelii*
4. *Dipcadi serotinum*
5. *Dipcadi serotinum*
6. *Dipcadi serotinum*
7. *Muscari comosum*
8. *Muscari comosum*

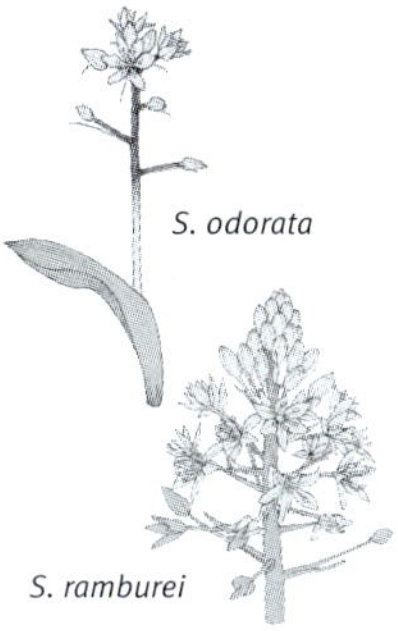

S. odorata

S. ramburei

Scilla monophyllos A small, bulbous perennial with stems to 30 cm and a **solitary leaf.** Leaf to 30 mm wide, linear-lanceolate and sheathing at the stem. Inflorescence a 4–12-flowered raceme becoming lax above; bracts to 7 mm long, exceeded by the flower stalks, at least below; flowers blue and star-like. Local on sandy ground and cliff tops. ***Scilla odorata*** is similar but with 1–few leaves, exceeding the flower stalk, **flowers scented**, and **lower bracts exceeding the flower stalks.** Common in seasonally wet places and around lake margins. ***Scilla ramburei*** is similar to *S. odorata* but with **racemes of up to 30 flowers** (not 10) and **bracts shorter than their flower stalks**; populations in the region have unusually long leaves and bracts less than half the length of the flower stalks. Sandy places.

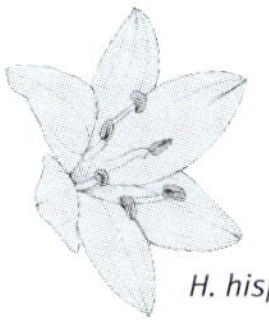

H. hispanica

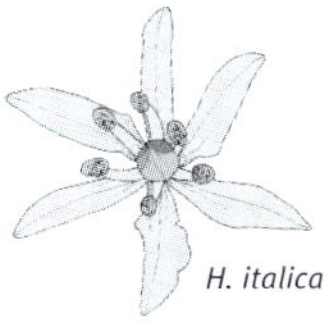

H. italica

Hyacinthoides hispanica SPANISH BLUEBELL Similar to *Scilla* species; but with flowers bell-shaped with spreading lobes (not star-like). Leaves 4–8, linear-lanceolate to lanceolate, and equalling or shorter than the flower stalk. Perianth bell-shaped and pale blue with spreading lobes; flowers nodding somewhat, not scented; anthers bright blue; **stamens inserted in the middle of the perianth**. Shady places and woods; local. ***Hyacinthoides italica*** (also described as *H. vicentina*) is similar but with narrow leaves to 8 mm wide, and flowers with **stamens inserted at the base of the perianth** (not the middle) with yellow anthers in some populations in the region. Woods and shady, rocky places.

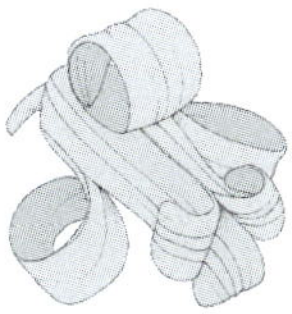

D. serotinum

Dipcadi serotinum A distinctive, spring-flowering hairless bulbous perennial. Leaves all basal and linear-lanceolate. Inflorescence an erect, lax raceme of up to 20 flowers facing more or less one way; **perianth brownish-orange**; bracts exceeding the flower stalks. Easily distinguishable by its flower colour. Locally common in rocky and sandy habitats; generally coastal.

Bellevalia Bulbous perennials with lax spikes of dull-coloured flowers; corolla with 6 deeply divided teeth; bell-shaped and tubular but not constricted into a throat. Fruit a 3-parted capsule.

Bellevalia hackelii A small, bulbous perennial to 30 cm tall with 2–5 leaves. Leaves held erect or recurved, equalling or slightly exceeding the flower stalks, linear and tapered towards the tip. Inflorescences cylindrical with 15–30 rather long-stalked flowers; perianth to 7 mm, bell-shaped, and **dark blue** when open, the lobes with a whitish margin. Common in dry, open and grassy habitats across the region.

Muscari GRAPE HYACINTHS Bulbous perennials similar to *Bellevalia* spp. but the corolla constricted at the mouth with 6 short teeth, the stamens hidden below the mouth and flowers borne in dense terminal clusters, with the uppermost flowers usually small and sterile.

M. comosum

Muscari comosum TASSEL HYACINTH A short-to-medium bulbous perennial to 50 cm tall. Leaves 3–6, linear, recurved, grooved and tapered gradually to the tip, to 40 cm long. Flowers borne in a lax raceme with **small, blue, sterile flowers held erect on long, slender blue stalks forming a tuft at the apex**; fertile flowers brownish-green, shortly tubular-bell-shaped with 6 outwardly curved short teeth. Common in open rocky and grassy habitats across the region.

M. neglectum

Muscari neglectum **COMMON GRAPE HYACINTH** A short, bulbous perennial. Leaves 3–6, linear, channelled and all basal. Inflorescence a short, dense raceme of touching ovoid blue flowers borne on spreading or recurved stalks; the uppermost flowers sterile and paler. Common in open rocky and grassy habitats across the region.

Agave Imposing trees and shrubs with fibrous, often enormous leaves in rosettes. Flowers cosexual and actinomorphic; stamens 6; ovary superior or inferior and 3-parted; seeds numerous.

A. attenuata

Agave americana **CENTURY PLANT** An enormous succulent to 9 m when in flower; plants take at least 10 years to flower, after which they die, but perennate by offsets. Leaves very large and succulent, to 2 m long and 30 cm wide, **with a stout black spine at the tip**; blue-green or variegated with yellow stripes. Flowers borne on a **tree-like inflorescence** that persists for years; flowers greenish-yellow, to 90 mm long. Fruit an oblong capsule. Common across the region, particularly in or near developed areas; cliffs and roadsides. ***Agave attenuata*** is similar but smaller, with less succulent leaves with a blue-white waxy bloom, and with flowers borne in an extremely dense, fox-tail-like inflorescence, nodding when mature. Planted and naturalised in coastal areas.

Ruscus Woody evergreen shrubs with stems flattened into leaf-like blades (cladodes) on which the flowers and fruits are borne directly.

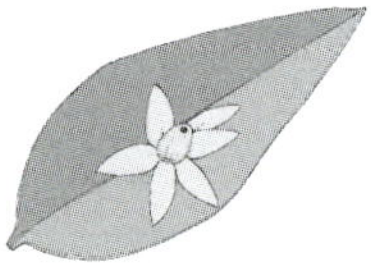

R. aculeatus

Ruscus aculeatus **BUTCHER'S BROOM** A dioecious dark green, evergreen shrub with ribbed stems, spiny and bushy with cladodes 10–40 mm long arranged in 2 rows, ending in a spiny tip. Flowers male or female on separate plants, inconspicuous, greenish-white, appearing to originate in the centre of the 'leaf'. Fruit a large red berry to 15 mm across. Common in woods at higher altitude.

Order: ARECALES

ARECACEAE

Palms with large, pinnately or palmately divided leaves. Flowers unisexual or cosexual, small and greenish, usually borne in spikes or panicles with papery sheaths; petals 3; stamens 6. Fruit a berry or drupe. A large, mainly tropical and subtropical family with several representatives planted in the region.

Chamaerops humilis **DWARF FAN PALM** A bushy, normally stemless dwarf palm with stems to 2 m in ungrazed areas. Leaves fan-like, tough and divided over $^{2}/_{3}$ their length into 20 pointed, narrowly lanceolate segments; stalk long and with a spine-toothed margin. Flowers borne in dense yellow panicles to 35 cm long, with sheathing bracts. Fruit oblong, to 30 mm, yellow or brown when ripe. Common on low *matos*, particularly near the coast.

P. canariensis

Phoenix canariensis **CANARY PALM** A very large palm to 20 m tall with a solitary trunk and crown; fronds to 6 m long with pinnate leaflets to 50 cm. Fruits to 25 mm long, orange ripening brown. Native to the Canary Islands but widely planted as an ornamental in towns and gardens in the region.

W. robusta

Washingtonia robusta A large palm to 20 m tall with an enormous trunk, the upper part clothed with the long-persistent withered leaves. Leaves large and fan-like. Native to Mexico; widely planted.

1. *Muscari neglectum*
2. *Agave americana*
3. *Agave americana*
4. *Ruscus aculeatus*
5. *Chamaerops humilis*
6. *Phoenix canariensis*

1

2

3

4

5

6

Order: POALES

JUNCACEAE

Erect, grass-like perennials with white, pith-filled stems. Differ from grasses and sedges in having actinomorphic flowers with 6 similar perianth segments and 6 stamens. Many species superficially similar; the anther length relative to the filaments is important and should be measured in late flower or in fruit.

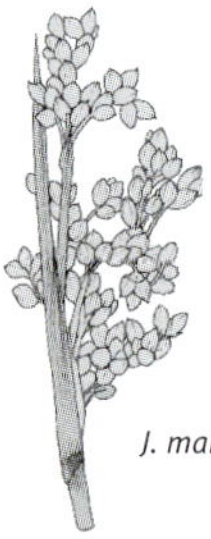

J. maritimus

Juncus maritimus SEA RUSH A tuft-forming rush with stems to 1 m, and 2 mm wide, and a short, creeping rhizome. Leaves almost basal, generally shorter than the stems, and **sharply pointed**. Inflorescence an interrupted panicle with 2–3 flower-heads; flowers straw-yellow and 6-parted with unequal perianth segments; anthers twice as long as the filaments. Bracts sharply pointed, the largest longer than the inflorescence. Capsules triangular-ovoid and pointed, to 3.5 mm long. Salt marshes, dunes and coastal grassland.

J. acutus

Juncus acutus SHARP RUSH Similar to *J. maritimus* but taller, to 2 m, leaves sharply pointed and longer than the stems. **Flowers reddish-brown**, borne in **dense, rounded inflorescences**, exceeded by their bracts; anthers up to 5-times longer than the filaments; perianth segments more or less equal. Capsule to 6 mm long. Coastal sands and salt marshes. ***Juncus striatus*** has **jointed, grooved leaves**, anthers 2–3-times as long as the filaments, and a **beaked capsule**. Damp habitats.

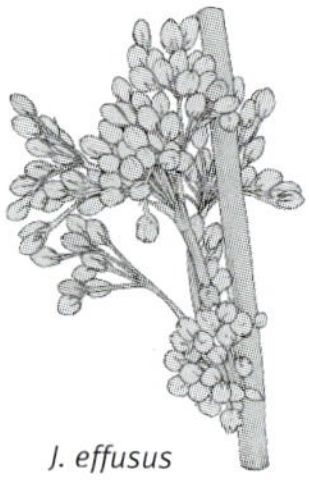

J. effusus

Juncus effusus SOFT RUSH A rush with soft-tipped leaves, and inflorescences of pale brown flowers with the **lowermost bract long and with a narrow sheath**, and **anthers scarcely longer than the filaments**; perianth segments more or less equal. Seasonally wet habitats and pool margins. ***Juncus bufonius*** has 3 stamens, with **anthers shorter or as long as the filaments** and unequal perianth segments; inner perianth segments longer than the capsule. Damp places. ***Juncus hybridus*** is very similar to *J. bufonius* but with 6 stamens (not 3), with anthers often to only half the length of the filaments, and inner perianth segments equalling or scarcely exceeding the capsule. Damp, saline habitats.

1. *Carex vulpina* PHOTO: TIM RICH
2. *Cyperus capitatus*
3. *Cyperus capitatus*

J. capitatus

Juncus capitatus DWARF RUSH A **dwarf, tufted annual** to 12 cm tall with all basal bristle-like leaves; several stiff stems with compact heads of flowers, overtopped by 1 or 2 bracts. Outer perianth segments greenish, later reddish with curved segments, exceeding the inner segments. Damp, sandy habitats.

CYPERACEAE

A large family of hairless, herbaceous sedges arising from a rhizomatous root system. Leaves grass-like but stems solid, and often triangular. Flowers wind-pollinated, borne on spikelets; stamens 2–3. Fruit a small, 2–3-sided nut.

Carex Sedges with unisexual flowers, each with a papery bract (glume), densely compacted in spikes; the upper usually male and the lower female.

C. vulpina C. divisa

Carex vulpina GREATER FOX SEDGE A stout, tufted sedge to 1 m with **sharply 3-angled to almost winged stems** with cylindrical, red-brown flower clusters to 80 mm. Flowers male above, female below with long-pointed glumes with green mid-veins and brown margins. Damp rocks and woods, streamsides; local. *Carex divisa* is a **creeping**, hairless perennial with **slender, wiry stems** and flattened leaves. Flowers male above, female below, in scarcely interrupted spikes; glumes short and translucent brown. Damp grassy and sandy places.

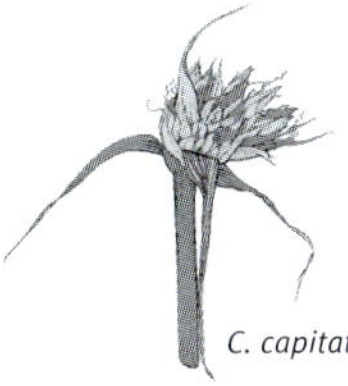

C. capitatus

Cyperus capitatus A small, tough, hairless, glaucous rhizomatous sedge with few, wiry leaves; stems solitary, to 30 cm tall. Leaves blue-green, becoming yellow with age, with inrolled margins. Bracts leaf-like, erect and exceeding the inflorescence; later brown and withered. Flowers borne in dense, brown, terminal spikelets. Common, sometimes abundant, on coastal dunes. *Cyperus eragrostis* is similar to *C. capitatus* but rather taller to 90 cm and more slender, and with greenish flowers borne in a simple to **compound umbel** of dense heads. Native to tropical America, naturalised in the region.

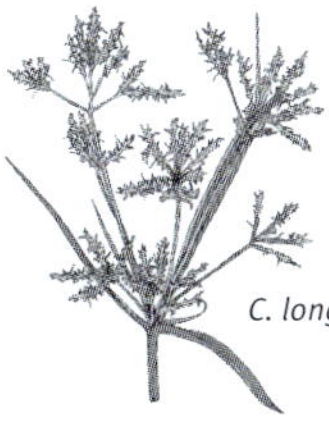

C. longus

Cyperus longus A rather robust sedge to 1.5 m tall with thick, far-spreading rhizomes. Leaves to 10 mm wide, shorter than or equalling the stems. Bracts 2–6, the outer exceeding the inflorescence. Inflorescence a simple or compound umbel with 8–12 **rays of brownish or reddish flowers**; stamens 3. A weed of damp places and pool margins.

C. laevigatus

Cyperus laevigatus A low to medium, tufted perennial with mat-forming rhizomes; stems erect and bunched or solitary, and rounded or triangular in section. Leaves few, and reduced. **Bracts 2, erect, and forming an apparent elongation of the stem**, and exceeding the inflorescence. Inflorescence a fascicle of 1–40 stemless spikelets with greenish-brown flowers; stamens 3. Local in marshes, stream margins, and other wet habitats.

E. palustris

Eleocharis palustris COMMON SPIKE-RUSH A hairless, aquatic perennial with creeping rhizomes with solitary (1st year) and later numerous, **tufted, leafless stems** to 60 cm, bearing cylindrical spikelets to 20 mm long. Stems reddish at the base with **leafless sheaths**; sheaths pale brown. Stems with approximately equal air canals in cross-section. Marshes and seasonally flooded habitats.

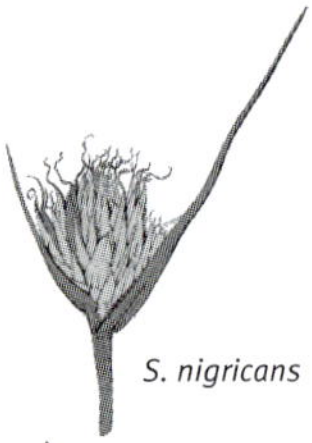
S. nigricans

Schoenus nigricans A small, densely tufted perennial with all leaves basal, and terminal inflorescences. Leaves half as long to slightly longer than the stems, dark grey-green, hard and wiry. Inflorescence with up to 10 rather flattened spikelets of 1–4 flowers; stamens and stigmas 3. Nuts 3-sided and creamy-white. Local on maritime sands and acid peat; common on Cape St. Vincent.

I. setacea

I. cernua

Isolepis setacea A small, slender, tufted, sedge-like herb with narrow stems to 30 cm long. Leaves few, and shorter than the stems. Inflorescence **much exceeded by the bract**, spikelets 2–3, up to 5 mm. Empty bracts (glumes) purple-brown with green midribs and translucent margins. Damp, marshy habitats. ***Isolepis cernua*** is similar but with **bracts shorter or scarcely exceeding the inflorescences**, which are borne on long, often drooping stems. Glumes greenish with a red-brown spot each side of the midrib. Damp coastal, sandy habitats.

POACEAE

Annual or perennial, tufted, often rhizomatous or creeping grasses. Leaves alternate, linear and sheathing the stem at the base; sheaths generally with a membranous ligule at the base. Inflorescences very variable, often spikes or panicles; flowers not brightly coloured, wind-pollinated and with 1, 3 or 6 stamens, and a pistil with normally 2 styles, enclosed within 2 bracts; the whole called a floret; florets arranged into spikelets with 2 empty bracts (glumes) at the base. A large and important family; it is beyond the scope of this book to describe all species in the region.

L. perenne

D. rigida

Lolium perenne RYE GRASS A hairless, wiry perennial grass to 50 cm. Stems smooth and slender, bent below. Leaves narrow, up to 3 mm wide, and **folded until mature**. Ligule to 1 mm, abruptly pointed. **Inflorescence simple**, to 15 cm long with compressed spikelets to 15 mm. Cultivated as a fodder crop.

Desmazeria rigida FERN-GRASS A low, hairless, **tufted, bluish annual** with erect stems. Leaves often purplish, fine-pointed, to 2 mm wide and flat or with inwardly rolled margins. Inflorescence a more or less 1-sided panicle to 80 mm long, often branched below, with sparse spikelets to 10 mm long. Coastal sands.

C. echinatus

Cynosurus echinatus ROUGH DOG'S-TAIL A short to tall hairless annual with erect or spreading stems and flat leaves to 5 mm wide. Inflorescence a dense, oblong panicle of **shiny green or purplish spikelets** to 12 mm long. Dry rocky and grassy scrub.

L. aurea

D. glomerata

B. madritensis

B. diandrus

B. hordeaceus

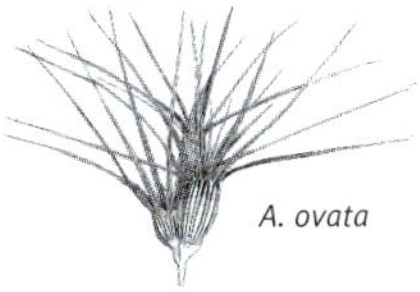

A. ovata

Lamarckia aurea **GOLDEN DOG'S-TAIL** A more or less hairless, low annual grass with tufted, erect stems. Leaves linear, only 1 mm wide with hairy margins; ligule membranous, to 10 mm long, pointed or blunt. Inflorescence dense, rather 'fluffy' and 1-sided with the outer spikelets sterile, greenish and later golden. Common on fallow and cultivated ground and roadsides; widespread.

Dactylis glomerata **COCK'S FOOT** A perennial, bluish clump-forming grass with erect or spreading stems to 1 m long. Leaves rough, with ligules 2–10 mm long. Inflorescence a tufted panicle of **laterally compressed spikelets borne in dense clusters on lateral branches**, often with prominent yellow stamens. Common in grassy habitats.

Bromus madritensis **COMPACT BROME** A short to medium tufted annual with erect stems. Leaves tapered, to 5 mm wide and **softly white-hairy**. Inflorescence a rather lax, erect, wedge-shaped panicle with short, bunched branches; spikelets to 60 mm long; flower-bract (**lemma**) with a long bristle-tip to 16 mm long; hairy or hairless. Common on fallow and cultivated ground. ***Bromus diandrus*** is similar to *B. madritensis* but with a long, **very lax, nodding inflorescence** (not wedge-shaped) with spreading spikelets. ***Bromus hordeaceus*** has an erect, rather **short and dense inflorescence** with **spikelets on short stalks**, spikelets which with a slightly inflated appearance.

Aegilops ovata A hairy or hairless tufted annual to 40 cm tall with flat, rough leaves and erect stems bearing small, ovate inflorescences to 20 mm long with a few long-bristled spikelets; the uppermost slender and sterile, the lower swollen; glumes thick-veined with 4 spreading, stiff bristles. Dry, uncultivated land.

1. *Schoenus nigricans*
2. *Lamarckia aurea*
3. *Aegilops ovata*

1. *Briza maxima*
2. *Hordeum murinum*
3. *Avena barbata*
4. *Lagurus ovatus*
5. *Ammophila arenaria*
6. *Elymus farctus* ligule
7. *Elymus farctus*

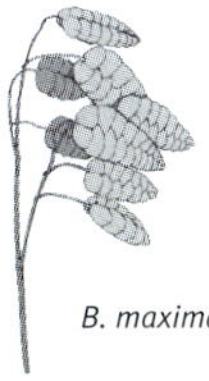

B. maxima

Briza maxima **LARGE QUAKING GRASS** A hairless, low annual grass with often solitary, erect stems. Leaves flat and linear, to 4 mm wide. Inflorescence a lax panicle, with up to 12 large, drooping **papery spikelets** appearing inflated, on slender stalks, green then purplish ripening pale brown, to 20 mm long. Common on fallow ground, roadsides and *matos*; widespread. ***Briza minor*** is similar but with **numerous (more than 12), smaller spikelets 3–7 mm long**. Open habitats.

H. murinum

Hordeum murinum **WALL BARLEY** A short annual with tufted, erect, smooth stems. Leaves linear, to 4 mm wide and hairy on both surfaces with shiny sheaths; ligule membranous and small, to 1 mm long. Inflorescence a more or less bilaterally symmetrical, bristly, dense, spike-like panicle with **awns** to 25 mm long; pale green or purplish. A common grass in a range of habitats across the region.

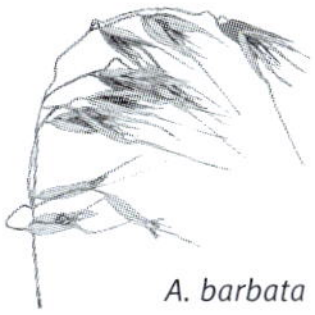

A. barbata

Avena barbata **BEARDED OAT** A medium to tall, erect grass with solitary or grouped stems. Leaves linear, to 15 mm wide and hairless or slightly hairy on the margin; ligule membranous, to 5 mm long. Inflorescence a 1-sided, very lax panicle with spikelets drooping on slender stalks; **lemma with 2 short bristles at the tip**. Fallow and cultivated ground and roadsides; widespread.

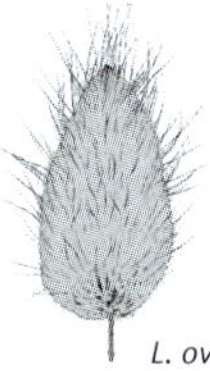

L. ovatus

Lagurus ovatus **HARE'S TAIL** A low to medium, softly hairy, grey-green annual. Leaves linear to lanceolate and flat; ligule hairy and membranous, to 3 mm long. Inflorescence distinctive; egg-shaped, dense, 'fluffy' soft and white, to 20 mm long; lemma semi-transparent and with a bristle-tip to 15 mm long. Common in sandy coastal environments and rocky slopes inland.

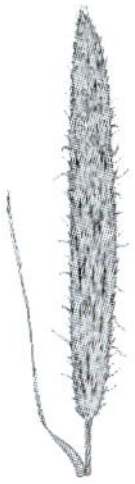

G. ventricosum G. phleoides

Gastridium ventricosum A small, annual grass to 50 cm tall with flat, hairless leaves, often withered when in flower; ligules to 3 mm long, and pointed. Flowers borne in green, **strongly erect** (at least at first), more or less bilaterally symmetrical and laterally compressed panicles to 12 cm long; **spikelets short, to 4 mm long with a single floret**. ***Gastridium laxum*** is arguably distinct from *G. ventricosum*, with lax panicles with florets to 3 mm long. ***Gastridium phleoides*** is native to Asia but naturalised as a **casual** weed in the region, and distinguished by its very dense panicles with **longer spikelets to 8 mm long**.

A. arenaria

Ammophila arenaria **MARRAM GRASS** A tough, clump-forming rhizomatous perennial grass with smooth stems. Leaves to 5 mm but appearing narrower (due to the inrolled margins); **ligules narrow and pointed**, to 25 mm long. Inflorescence a slender, spike-like panicle with dense, erect spikelets to 16 mm long; lemma with a very short, stiff bristle-tip with a ring of hairs at the base. Very common to dominant on coastal dunes across the region.

E. farctus

Elymus farctus A tough, clump-forming, rhizomatous perennial grass, rather similar to, and often co-occurring with *Ammophila arenaria*. Leaves inrolled at the margins and minutely hairy above, hairless beneath; **ligules short**. Inflorescence a slender spike with dense, erect, virtually stalkless, laterally compressed and hairless spikelets; **lemma without a bristle-tip**. Common on coastal dunes across the region.

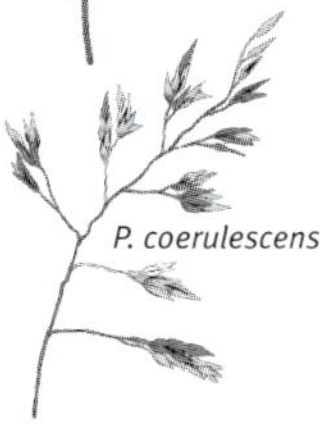

Piptatherum miliaceum A tall, lax grass to 1.5 m with erect or ascending, smooth, hairless stems. Leaves to 5 mm wide, rough above; ligule to just 1 mm long and abruptly cut at the tip. Flowers borne in light green to straw-coloured panicles to 40 cm long; 1-sided, drooping and lax, with several (to 20) branches at each node along the stalk, those below without spikelets, upper branches with **short spikelets to just 4 mm long**. Dry open habitats among shrubs and other vegetation. ***Piptatherum coerulescens*** is similar but **smaller**, 30–70 cm tall with narrower leaves to 2.5 mm wide only, rough on both surfaces, and a **longer ligule to 11 mm long**. Dry, stony habitats.

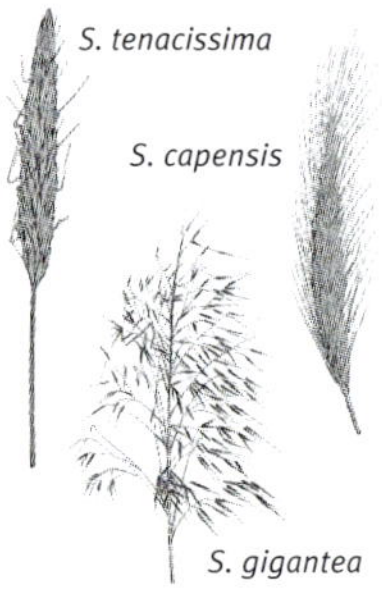

Stipa tenacissima **ESPARTO GRASS** A **very large**, tough, tufted perennial grass to 1.5 m tall. Leaves inrolled with overlapping margins; ligule short, hairy at the margin, with a **feathery tip**. Inflorescence a long, **dense and compact, spike-like panicle** to 35 cm long; spikelets 1-flowered, the lemma with a notched tip and long bristle to 60 mm long. Local but forming extensive colonies in open, often sandy habitats. ***Stipa gigantea*** **(SPANISH OAT-GRASS)** is very similar to *S. tenacissima* but with **very lax panicles** of flowers. Similar habitats, subdominant on dunes on Cape St. Vincent. ***Stipa capensis*** is a smaller **annual or biennial** to just 50 cm, with dense, spike-like panicles, and lemma with a very long bristle to 12 cm long.

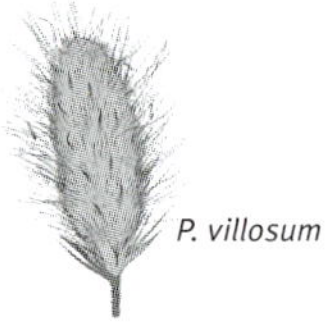

Pennisetum villosum A large, pampas-grass-like, laxly clump-forming perennial grass to 1 m tall. Leaves to 70 cm long and rigid. Inflorescences terminal and rather like those of *Lagurus ovatus* but larger, 'fluffy' and white and broadly cylindrical, borne in larger numbers terminally, drooping on slender stems; spikelets to 14 mm. A casual weed, naturalised in dry habitats.

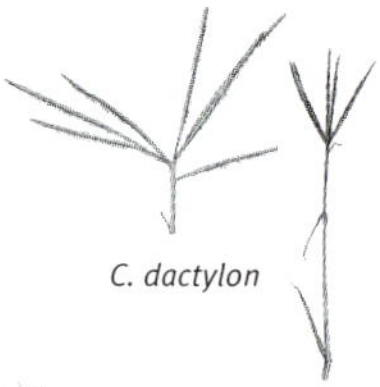

Cynodon dactylon **BERMUDA GRASS** A patch-forming low to short perennial. Leaves linear and flat, hairless or hairy along the margins. Inflorescence distinctive and star-like: **3–6 spikes outwardly spreading from a single central axis**; spikelets to 3 mm long, stalkless. Native to tropical Africa but a commonly naturalised weed in a range of disturbed habitats.

Hyparrhenia hirta A tufted, medium to tall perennial grass with smooth stems. Leaves linear, to 3 mm wide and more or less hairless. Inflorescence a panicle of **paired racemes**, each enclosed in leaf-like bracts; lemma to 4.5 mm long with a stout bristle-tip to 20 mm long, twisted and hairy below. Fairly frequent in dry, rocky habitats.

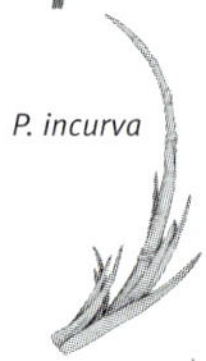

Parapholis incurva **CURVED SEA HARD-GRASS** A short, tufted annual with **spreading, curved stems**. Leaves linear and pointed, to 2 mm wide, rough above and along the margins with reddish sheaths. Inflorescence a cylindrical spike, **strongly curved and jointed** with spikelets appressed to the stem. Saline coastal habitats.

Sorghum halepense **JOHNSON GRASS** A **large**, erect perennial grass to 2.5 m tall with **stems silkily hairy at the nodes**. Leaves hairless with a rough margin. Inflorescence a large terminal panicle to 30 cm long; spikelets to 5.5 mm long; lemma notched at the apex and with an awn to 12 mm long. Locally common in damp, disturbed habitats.

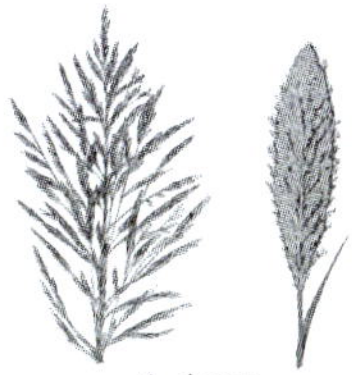

A. donax

P. australis

Arundo donax **GIANT REED** An **extremely large** and robust, rhizomatous perennial grass with **bamboo-like stems** to over 6 m. Leaves grey-green, to 60 mm wide. Inflorescence a large panicle to 60 cm long; silky and silvery, with spikelets to 20 mm long, each with usually **3 florets**; lemma notched at the apex with a short bristle-tip. Common in damp places, sometimes planted in long banks along roadsides in the region.

Phragmites australis A large, reed-like grass, rather similar to *A. donax*; a rhizomatous perennial with tall, rather thin stems to 3 m which do not overwinter. Leaves up to 50 cm long and 50 mm broad, grey-green and tapered to the tip; sheaths smooth and hairless, surrounding the leaf nodes. Flowers borne in drooping, 1-sided, more or less cylindrical, bunched greenish or purplish panicles, with spikelets to 16 mm long, **each with up to 10 florets**.

1. *Stipa gigantea*
2. *Stipa gigantea*
3. *Cynodon dactylon*
4. *Arundo donax*
5. *Arundo donax*

Corynephorus canescens A low, **tufted, bluish-grey grass** with rigid bristle-like leaves, of more or less similar length. Spikelet purple and white with 2 florets, compressed, in a compound panicle, exceeded by the single-veined, membranous glumes; anthers orange or purple, much longer than broad. Common on coastal dunes.

Cortaderia selloana PAMPAS GRASS A very tall, dense grass to 3 m tall. Leaves long and slender, to 2 m long and 10 mm broad, with very sharp edges. Feathery flowers are borne in dense plumes which overtop the leaves. Native to South America and widely planted in towns and gardens.

Secale cereale RYE is similar to *Hordeum*, but a cultivated crop grown on poor, shale substrates in the region.

Oryza sativa RICE is another cultivated grass that is grown in wet fields along the western coast belt of the Algarve.

EUDICOTS **[Eudicotyledons]:** 'True dicots', the largest monophyletic group of angiosperms and major component of the former 'dicotyledons'. These plants (like basal angiosperms) typically have two cotyledons (seed leaves), leaves with a network of veins radiating from a central main vein (rather than parallel veins), flower parts in multiples of two, five, or seven, and a ring of vascular cambium in the stem. In contrast to most monocots, many eudicots undergo secondary growth, forming trees and shrubs.

Order: RANUNCULALES

RANUNCULACEAE

The buttercup family contains herbs and climbers with alternate, often dissected leaves. Flowers often showy, normally with 5 sepals and petals; stamens numerous. Fruit a collection of achenes.

Anemone palmata **YELLOW ANEMONE** A low, hairy perennial with a tuberous rhizome. Basal leaves more or less circular with 3–5 shallow, toothed lobes; upper leaves united at the base. Flowers bright yellow, to 35 mm across with up to 15 petals. Common in rocky habitats and on the *matos*.

Ranunculus ficaria **LESSER CELANDINE** A variable, hairless perennial with **heart-shaped**, fleshy, dark green leaves. Flowers shiny and yellow, turning white on ageing with 8–12 narrow petals and 3 sepals. Very common in damp, grassy places.

R. bullatus

Ranunculus bullatus A low, hairy perennial with leaves all basal, stalked, ovate to rounded and toothed but not lobed. Flowers solitary on leafless stalks, yellow, to 26 mm across with 5–12 petals, and **sweetly scented**. Local on rocky slopes.

Ranunculus muricatus A short, usually hairless annual with **kidney-shaped**, 3–7-lobed lower leaves; upper leaves with up to 5 lobes or occasionally entire. Flowers yellow, to 16 mm across. Fruit rounded, with achenes with a broad wing with short spines. Fairly common in ditches and in other damp, grassy places.

Ranunculus bupleuroides A perennial with entire, broad, petiolate blue-green, oblong-lanceolate leaves which are often inwardly curved, with **distinct paler venation**. Flowers yellow, with 5–6 petals, each with a basal nectary. Local in the Monchique area.

Ranunculus aquatilis An annual or perennial aquatic herb with floating and submerged leaves, the latter sometimes absent; floating leaves deeply divided into 3–7 **straight-sided segments with teeth at the tips. Flowers up to 18 mm in diameter**. Fruit stalk up to 50 mm long. Locally common in suitable habitats; occurs in still or slow-moving water. ***Ranunculus peltatus*** is similar but with **leaf segments with rounded tips**, and **larger flowers** to 30 mm across, and a fruit stalk to 15 cm long. Locally common in aquatic habitats, but varies in abundance with seasonal rainfall.

---

1. *Corynephoros canescens*
2. *Anemone palmata*
3. *Ranunculus ficaria*
4. *Ranunculus muricatus*
5. *Ranunculus peltatus*

D. staphisagria

Nigella hispanica **Spanish love-in-a-mist** A short, hairless annual with leaves alternate, and finely divided into **linear segments**. Flowers to 70 mm in diameter, bright blue, with reddish anthers. Follicles of the fruit fused for most of their length, and densely glandular. Frequent in a range of habitats.

Delphinium staphisagria A medium to tall hairy annual or biennial with 5–7 large, oblong, tri-lobed segments. Flowers deep blue, large to 25 mm across, borne in long racemes. Flowers with short, blunt, down-turned spurs that are shorter than the petals. Local on rocky slopes and on the *matos*.

C. ambigua

Consolida ambigua **larkspur** Similar to *Delphinium* spp. but with only 2 inner perianth segments, which together form a spur. **Leaves feathery** with many thread-like segments. An annual to 1 m. Flowers stalked, deep blue in a long and many-flowered inflorescence. In fields and on dry, rocky slopes. ***Consolida orientalis*** is similar but stickily hairy, and with short-stalked flowers borne in a dense raceme. On cultivated land and waste ground.

PAPAVERACEAE

Annuals or perennials with milky or coloured sap. Leaves 1–2-times pinnate. Flowers solitary or in lax clusters. Sepals 2, petals 4, crumpled looking. Stamens numerous. Fruit a dehiscent capsule.

P. somniferum

Papaver somniferum **opium poppy** A vigorous, erect, **glaucous annual** with pinnately divided, ovate leaves. Lower leaves with a short petiole, the upper leaves clasping the stem. Flowers large, to 90 mm across, pale purple with a dark centre, anthers yellow. Capsule hairless. A relic of cultivation and widely naturalised in waste places. **Subsp. *setigerum*** is similar but shorter, often only to 15 cm, and with sparse long, fine bristles. The leaves are often more deeply lobed, and **end in a bristle**. Flowers white to pale purple, darker in the centre. Probably native on coastal sands.

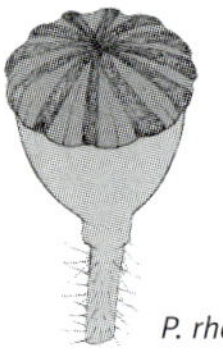

P. rhoeas

Papaver rhoeas **common poppy** An erect, bristly annual to 50 cm tall. Leaves pinnately lobed, to 15 cm long with pointed segments, often twice pinnate. Flowers solitary on long stalks with long, bristly hairs; petals crimson with or without a dark centre, anthers bluish. Capsule more or less **rounded, and hairless**. Common on cultivated and disturbed ground throughout.

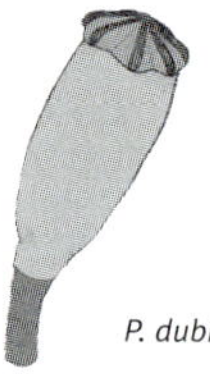

P. dubium

Papaver dubium **long-headed poppy** Superficially very similar to *P. rhoeas* but with appressed hairs on the upper parts of the stem, the leaf segments blunt, not pointed, flowers a paler or more orange-red, usually without a dark centre, and the capsule **oblong**, somewhat widened towards the apex. Common on cultivated and disturbed ground throughout.

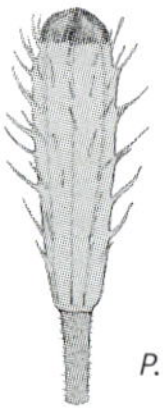

P. argemone

Papaver argemone **prickly poppy** A short to medium, bristly annual with appressed hairs on the stem, and pinnately divided leaves. Flowers pale scarlet, 20–60 mm across, often with a dark centre, and the **petals not overlapping**. Capsule oblong-cylindrical, ribbed, and sparsely **bristly**. Common on sandy waste ground, often coastal. ***Papaver hybridum*** is similar to *P. argemone* but with darker crimson-red flowers, and an ovoid to globose capsule that is covered in pale, stiff bristles. Habitats similar to those of *P. argemone*.

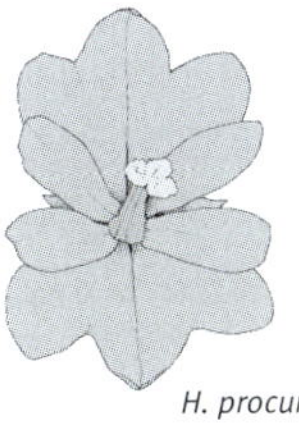

H. procumbens

H. imberbe

Glaucium flavum **YELLOW HORNED-POPPY** A glaucous, branched biennial to perennial, producing a yellowish latex. Leaves oblong, wavy and pinnately lobed, the upper leaves clasping the stem. Flowers yellow, to 75 mm across. Fruit narrowly cylindrical and **very long**, up to 30 cm; curved and hairless. On coastal sands and shingle; local.

Hypecoum procumbens A delicate, low to short, hairless, glaucous annual with wide-spreading stems to 15 cm. Leaves twice pinnately lobed, segments lanceolate or linear. **Bracts leaf-like.** Flowers borne in small, branched clusters, to 15 mm across, with 4 yellow petals: 2 large 3-lobed petals and 2 small lateral petals. Capsule erect and jointed. Local on maritime sands. ***Hypecoum imberbe*** is similar but has more erect with **linear bracts**. Flowers orange-yellow with more or less **evenly lobed large outer petals**. Fruits scarcely jointed. On cultivated land and waste places.

1. *Papaver somniferum*
2. *Papaver somniferum* fruit
3. *Papaver somniferum* subsp. *setigerum*
4. *Papaver dubium*
5. *Papaver dubium* fruit
6. *Papaver dubium*
7. *Papaver hybridum*

Fumaria Delicate trailing or scrambling herbs with much-divided leaves, and distinctive tubular 2-lipped flowers with 2 small sepals, 2 outer petals and 2 narrower inner petals. Fruit normally a rounded 1-seeded nutlet.

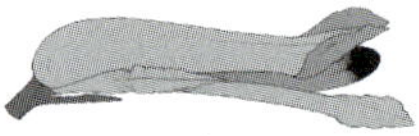
F. agraria

Fumaria agraria A short annual with broad, flat, ovate leaf segments. Flowers 2-lipped, pale pink, only the inner petals with dark purple tips, and the lower petal spreading at the margin, borne in a raceme of up to 20 flowers. Racemes longer than their stalks.

1. *Fumaria capreolata*
2. *Drosophyllum lusitanicum*
3. *Drosophyllum lusitanicum* PHOTO: FINN RASMUSSEN
4. *Drosophyllum lusitanicum* PHOTO: FINN RASMUSSEN

1

3

2

4

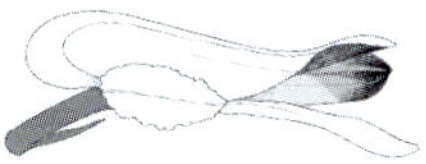

F. capreolata

Fumaria capreolata **WHITE RAMPING-FUMITORY** Rather similar to *F. agraria*; a tall, hairless, blue-green, scrambling annual without tendrils. Leaves wedge-shaped with blunt lobes. Flowers 2-lipped, the **upper petal compressed with upturned margins not concealing the keel,** creamy white, often tinged with pink, and **tipped with reddish-black**, borne in a cylindrical raceme of up to 20 flowers. Fruit stalk strongly curved. **Racemes shorter than their stalks.** Very common on disturbed ground throughout. ***Fumaria bicolor*** is similar to *F. capreolata* but with **flowers gradually turning pink**, tipped with dark purple, in a short 8–12-flowered raceme. ***Fumaria muralis*** is similar to both species but with the upper petal bearing **broad wings concealing the keel**, and **racemes as long as, or longer than their stalks**.

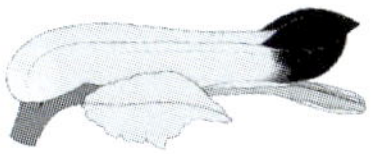

F. muralis

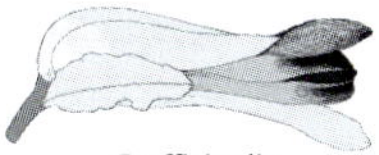

F. officinalis

Fumaria officinalis **COMMON FUMITORY** A very delicate hairless, blue-green, scrambling annual with broad, flat, ovate to lanceolate leaf segments. **Flowers many (up to 40), mauve**, tipped with blackish-purple on the wings of the upper petal and at the apex of the inner petals, to 9 mm long, borne in a **raceme longer than the short stalk**. Common on disturbed ground.

Platycapnos spicata A grey-green annual to 30 cm tall with twice pinnately lobed leaves. Flowers tiny, borne in dense rounded heads, superficially clover-like; petals creamy to pink, tipped red. Fruit ovate. Local on disturbed ground.

Order: PROTEALES

PLATANACEAE PLANE TREE FAMILY

A small family of trees with peeling bark and separate male and female flowers.

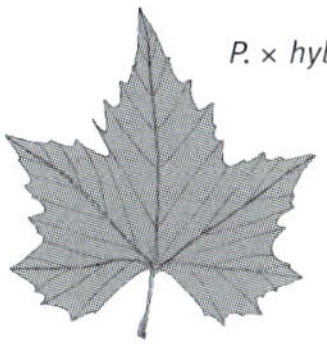

P. × hybrida

Platanus × hybrida **PLANE TREE** A large, deciduous tree to 35 m with bark peeling in flakes, and 5–7-lobed palmate leaves. Flowers borne in dense globular heads, often in pairs. Fruit globular and pendent. Not native (origin uncertain) but widely planted on roadsides in the area.

Order: CARYOPHYLLALES

DROSOPHYLLACEAE

Previously placed in the Droseraceae, but now established to be phylogenetically distinct. Plants carnivorous by virtue of the sticky hairs covering their leaves.

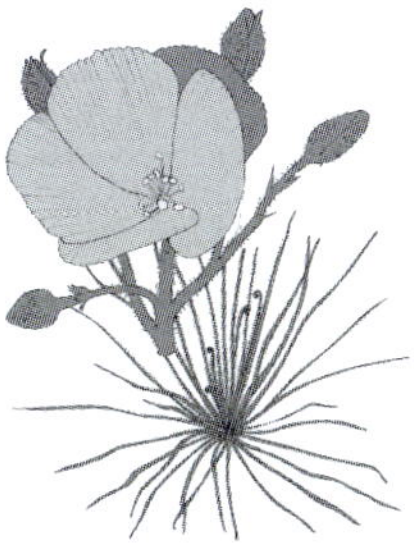

D. lusitanicum

Drosophyllum lusitanicum **PORTUGUESE SUNDEW** A distinctive slender, **very sticky-glandular perennial**, with ascending stems and leaves in a dense rosette. Leaves linear and tapering, covered in red-tipped glandular hairs. Flowers yellow, with 5 petals and borne in lax clusters on a scape to 30 cm long. Very rare and much declined through collection by carnivorous plant enthusiasts. Dry rocky habitats. The previous stronghold for this species near Algos is threatened with development and the plant may be extinct there.

TAMARICACEAE

A taxonomically difficult family of shrubs or small trees with simple, alternate, small ,scale-like leaves. Flowers borne in catkin-like spikes, 4–5-parted with 4 or more stamens. Fruit a capsule with hairy, wind-borne seeds.

Tamarix africana A feathery, more or less hairless tree to 6 m with blackish bark. Leaves small, to 4 mm long, pointed and growing close to the stem. Flowers white or pale pink in racemes to 60 mm long, borne on the previous year's wood alongside the new leaves; petals 5, to 3 mm long and usually **persistent in fruit**. Bracts hairy. Local in coastal marshes and near streams. ***Tamarix canariensis*** is very similar but minutely hairy with reddish-brown bark, with small, pale pink petals 1–2 mm that **fall in fruit**. Habitats similar to those of *T. africana*.

FRANKENIACEAE

Dwarf shrubs with opposite, entire leaves and no stipules. Flowers usually cosexual with 5 partly fused sepals and 5 free petals; stamens 6. Fruit a small capsule.

F. laevis

F. pulverulenta

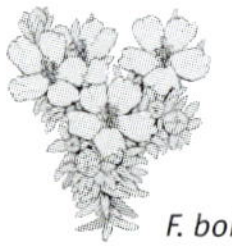

F. boissieri

Frankenia laevis SEA HEATH A low, prostrate, mat-forming, often minutely hairy perennial with tiny white-encrusted leaves to 5 mm long; woody at the base. Flowers pale to deep pink borne on short lateral or terminal and crowded spikes. Common on sea cliffs in the Cape St. Vincent area. ***Frankenia pulverulenta*** is similar but is annual, with leaves without down-turned margins and not white-encrusted. Maritime sands and shingle. ***Frankenia boissieri*** is an **erect (not mat-forming) perennial** to 30 cm with leaves to 7 mm, and purplish flowers borne in dense terminal clusters. Saline maritime habitats.

PLUMBAGINACEAE

Perennial, characteristically thrift-like plants with untoothed, sometimes grass-like leaves. Flowers borne in lax or tight clusters, 5-parted with papery calyx lobes that persist in fruit; petals fused in the lower part to form a tube. Fruit a capsule.

A. gaditana

Armeria pungens A thrift-like, compact, dwarf shrub with a woody stock. Leaves linear, to 10 cm long, rather grass-like and tufted. Flowers white or pale pink in heads to 30 mm across borne on stalks to 25 cm long. Common on coastal sands in the region. ***Armeria macrophylla*** is similar but with longer, **thread-like leaves** to 25 cm long. Scrub and sandy or stony places. ***Armeria gaditana*** is similar to *A. pungens* but with much broader, strap-like leaves to 35 mm wide, and deep pink flowers. Only in estuarine meadows. ***Armeria velutina*** is a **herb covered in dense, soft, silky hairs**. Scrub on dry acid soils in the east of the region. ***Armeria rouyana*** is recorded from the southern Alentejo region on sandy scrub and may also occur in the Algarve. It is characterised by the leaves that lack a sharp point (awn) and bracts around the flowers (**involucral bracts**); bracts leathery, pointed, and with a broad, membranous margin at the apex.

Limonium SEA LAVENDERS Perennial herbs with simple, leathery leaves forming basal rosettes and flowers in corymbs. The flowers persist and become papery in fruit.

L. sinuatum

Limonium sinuatum WINGED SEA LAVENDER A short to medium bristly perennial that could perhaps be confused with members of the Boraginaceae. Leaves pinnately divided with 4–7 pairs of rounded lobes, to 15 cm long in a flat, basal rosette. Stems with 3–4 undulate wings. Flowers yellow-white or pink, later purple; **calyx conspicuous, blue-purple and with a papery margin**. Saline coastal areas.

1. *Tamarix africana*
2. *Tamarix africana*
3. *Frankenia laevis*
4. *Armeria pungens*
5. *Armeria pungens*

L. ferulaceum

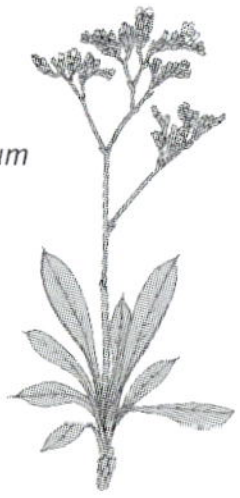

L. vulgare

Limonium ferulaceum A densely branched dwarf shrub with dense, brush-like tufts of flowering and non-flowering branches, to 40 cm tall. Leaves elliptical, unlobed, and notably, **absent when in flower**. Flowers pink, to 6 mm, borne at the ends of the branches. Rocky and sandy habitats and saline, marshy areas. ***Limonium diffusum*** is similar but with **some leaves present when in flower**, and flowers along the branches (not restricted to the tips). Flowers pale violet. Coastal sands and salt marshes.

Limonium vulgare COMMON SEA LAVENDER A short to medium, hairless perennial with bluish, spoon-shaped, pinnately veined **leaves borne in a lax rosette**. Flowers crowded, reddish to lilac, to 8 mm on stems branched above and spreading. Common to subdominant on mud flats and marshes.

1. *Limonium vulgare*
2. *Limonium vulgare*
3. *Limonium ovalifolium*
4. *Polygonum maritimum*
5. *Rumex bucephalophorus*

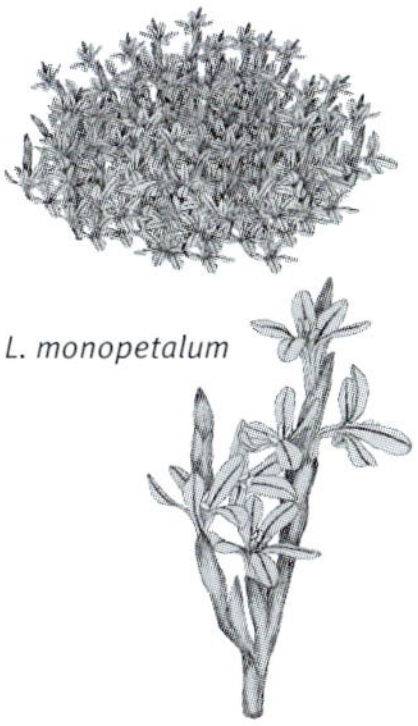

L. monopetalum

Limonium oleifolium A smooth, hairless perennial to 50 cm with rather flat, ovate, **prominently single-veined leaves** on long basal branches. Non-flowering branches numerous, flowering branches with pink-purple flowers to 8 mm across. Coastal rocks and sands.

Limonium ovalifolium is similar but forms **dense, cushion-like clusters of leaves**; leaves green, often tinged purple and rather concave, with indistinct venation. Flowers lilac, borne on spreading branches. Local in the region; common on the cliff tops of Cape St. Vincent.

Limoniastrum monopetalum LIMONIASTRUM A small, fleshy shrub to 50 cm with silvery-green, fleshy spoon-shaped leaves covered in white scales, sheathing the stem at the base. Flowers bright pink and conspicuous, later violet, to 16 mm borne in loosely branched spikes; corolla with 5 spreading, oval petals. Maritime sands and salt marshes.

POLYGONACEAE

Herbs or small shrubs without latex, with alternate leaves and stipules that form a membranous sheath around the stem. Flowers hermaphrodite or unisexual, often small and greenish or reddish.

Polygonum maritimum SEA KNOTGRASS A prostrate or ascending branched perennial with a woody stock. Leaves narrowly elliptical, glaucous and sessile, with down-turned margins. Stipules silvery, reddish at the base and strongly veined. Flowers white or pink, 5-lobed and to 4 mm across, solitary or in groups at the nodes. Common on maritime sands and shingle.

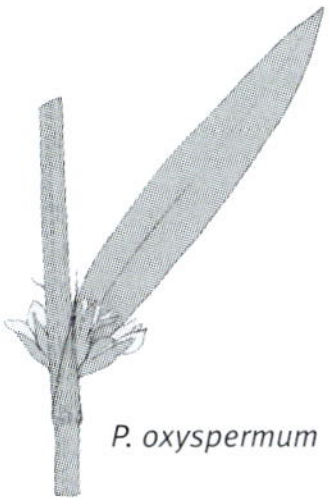

P. oxyspermum

Polygonum oxyspermum An annual or perennial with stems to 1 m, rather stout and with a woody stock. **Leaves elliptical to linear-lanceolate**, margins scarcely rolled downwards at the margins. Stipules much shorter than the **internodes**, transparent with 4–6 simple veins. Flowers in clusters of 2–6 (rarely solitary), bracts leaf-like. Nut exceeding the perianth. Locally common on maritime sands and shingle.

R. tingitanus

Rumex tingitanus TANGIER DOCK A rhizomatous perennial to 50 cm, branched at the base with greyish, **oval-spear-shaped leaves** that are undulate, sometimes lobed, often tapered towards the apex. Inflorescence with large, persistent sheaths. Fruit valves much longer than the nut. Locally common on maritime sands and shingle.

Rumex intermedius Perennial with stems to 60 cm, often clumped. **Leaves linear**, 10 cm long and 10 mm wide, with narrow divergent lobes and an **undulate margin**. Fruit valves longer than the nut. Locally common on maritime sands and shingle.

R. intermedius

Rumex bucephalophorus A variable low to medium, erect annual to 40 cm, branched or not. Leaves normally small, to 20 mm long, lanceolate, ovate or spoon-shaped, stalked, and greyish-green. Flowers borne on variously sized pedicels, very small, red and in clusters of 2–3 in the leaf axils, forming a dense spike. Fruit on recurved stalks; 3 of the perianth segments enlarged with straight or hooked teeth on each side. Very common on waste ground, maritime sands and sea cliffs throughout.

E. spinosa

Emex spinosa **EMEX** A hairless, short, somewhat fleshy annual with sprawling stems. Leaves oval, heart-shaped at the base and stalked. Male flowers stalked in terminal clusters, female flowers sessile at the base. Fruit a spiny nut. Locally common on maritime sands and disturbed ground.

CARYOPHYLLACEAE

Herbs with opposite leaves and cosexual, actinomorphic flowers with 4–5 free or fused sepals and 4–5 petals (absent in some species), 8–10 stamens or fewer, and 2–5 stigmas. Fruit a capsule.

A. montana

Arenaria montana **MOUNTAIN SANDWORT** A robust, mat-forming, shortly pubescent perennial with spreading stems. Leaves opposite, entire, oblong-lanceolate or linear, 1-veined. Flowers 5-petalled, solitary or in 2–11-flowered clusters, pedicels longer than sepals, petals twice as long as sepals, and white, yellow-centred. Capsules ovoid, more or less shorter than sepals. Common on dry, rocky slopes and *matos*.

C. glomeratum

Cerastium siculum A small annual to 12 cm covered in minute glandular hairs. Leaves ovate-elliptical, to 10 mm long, hairy. Flowers with white, **deeply notched petals that are shorter than the sepals**. Pedicels shorter than the sepals. Rocky hillsides and scrub.

Cerastium glomeratum A small to medium annual to 30 cm high (often much less), covered in sticky glandular hairs. Leaves ovate-elliptical, to 20 mm long, hairy. White **flowers borne in dense clusters**, the **sepals hairy to the apex**, petals more or less equalling the length of the sepals. Very common on bare ground and in dry, grassy places. ***Cerastium fontanum*** is similar, but a non-sticky, short-lived **perennial, with short, non-flowering basal shoots**, sessile leaves, flowers not in dense clusters, petals scarcely longer than the sepals; habitats similar to those of *C. glomeratum*.

P. argentea

P. echinulata

P. capitata

Paronychia argentea A branched, mat-forming perennial to 30 cm long. Leaves ovate-lanceolate, greyish and in opposite pairs, bristle-tipped and almost hairless. Stipules membranous, and shorter than the leaves. Flowers in rounded clusters; tiny, with 5 equal sepals, dwarfed by the membranous, silvery bracts. Calyx lobes with transparent margins. Common in maritime sandy places throughout. ***Paronychia echinulata*** is similar but an **annual**, spreading to erect, and with much less conspicuous bracts; in habitats similar to those of *P. argentea*. ***Paronychia capitata*** is similar to *P. argentea* but with hairier leaves without bristle-tips, equalling the length of the stipules, and denser flower-heads and unequal, orange sepals recurved at the tips.

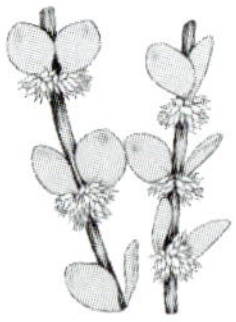

I. verticillatum

Illecebrum verticillatum **CORAL NECKLACE** A distinctive, low, hairless annual superficially similar to *Paronychia* spp. Stems creeping, rooting at the base. Leaves 2–5 mm, opposite and with small stipules, Flowers borne in 2 clusters at each node, 4–6-flowered with silvery bracts, looking like white 'pseudo-whorls'. Local in damp, sandy and gravelly places.

H. maritima

Herniaria maritima An inconspicuous, low, prostrate, spreading and more or less mat-forming annual with older stems woody at the base, **younger stems hairy on one side only**. Leaves tiny, to 5 mm, broadly elliptical and with appressed hairs on both surfaces. Flowers tiny, to 2 mm across, 5-parted. Rare, on maritime sands. ***Herniaria hirsuta*** is bright pale green but with grey-white younger shoots which are clothed in **dense, straight, white, spreading hairs**, and flowers with 2–5 stamens, borne in roundish clusters in the leaf axils; **sepals bristle-tipped**. Maritime sands.

1. *Cerastium fontanum*
2. *Cerastium glomeratum*
3. *Cerastium glomeratum*
4. *Arenaria montana*
5. *Paronychia argentea*
6. *Illecebrum verticillatum*
7. *Herniaria maritima*

1. *Polycarpon tetraphyllum*
2. *Loeflingia hispanica*
3. *Sagina apetala*
4. *Sagina procumbens*
5. *Spergularia media*
6. *Spergularia marina*
7. *Spergularia bocconii*

P. tetraphyllum

Polycarpon tetraphyllum **FOUR-LEAVED ALLSEED** A tiny, hairless micro-annual without a woody stock, to about 20 mm high. **Leaves mostly in whorls of 4**, oval, green or purple. Flowers in branched clusters, white, tiny to 2 mm across with notched petals shorter than the sepals. Common on dunes and in bare, sandy places inland.

Corrigiola litoralis **STRAPWORT** A low, flat, **greyish annual** with spreading or prostrate stems bearing rather fleshy, blunt, entire leaves with papery stipules at their bases. Flowers white, tiny, and borne in dense, rounded clusters along the stems; sepals green with white margins; anthers purplish. Local on damp coastal sands.

L. hispanica

Loeflingia hispanica **LOEFLINGIA** A low, glandular-hairy, reddish or purplish, and much-branched annual with linear, long-pointed leaves fused at the base. Flowers greenish, tiny to 3 mm across in branched spikes. Sepals and petals 5, stamens 3. Common in dry, sandy habitats.

S. apetala

Sagina apetala A small annual herb with slender, erect stems. Leaves linear and tapered at the tip. Flowers solitary, small with ovate sepals that are often hooded; petals minute and falling early. Common in sandy places throughout. ***Sagina procumbens*** is a similar **mat-forming perennial** with a short, non-flowering main stem bearing a central, dense leaf rosette and numerous lateral stems ascending from rooting bases. Common in sandy and urban waste places.

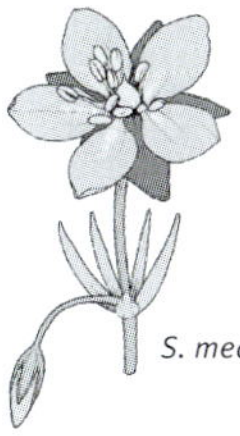

S. media

Spergularia media **GREATER SEA SPURREY** A low-short almost hairless perennial with fleshy leaves ending in an abrupt point. Stipules broadly triangular. Flowers to 12 mm across with white or pink **petals that equal or exceed the sepals**. Capsule much exceeding the calyx, seeds dark brown. Common on sandy shores and in salt marshes.

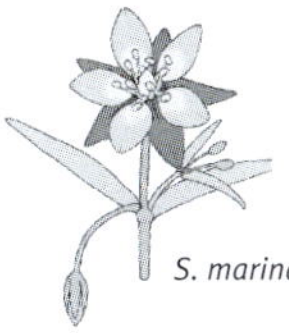

S. marina

Spergularia marina **LESSER SEA SPURREY** Similar to *S. media* but with very fleshy leaves, normally a slender (not woody) stock, and short stipules that form a sheath. Inflorescence little-branched; petals pink above and white below, **not exceeding the sepals**. Capsule exceeding the calyx; seeds light brown. Common on sandy shores and in salt marshes. ***Spergularia tangerina*** is similar but **smaller**, with very slender stems to just 50 mm, and a condensed inflorescence, flowers small to 3 mm across. Local, on saline soils.

S. bocconii

Spergularia bocconii A slender annual or biennial, **densely hairy on the inflorescence**. Leaves not in dense clusters, stipules triangular. Petals pink with a white base, equalling or shorter than the sepals. Capsule shorter than the calyx; seeds grey-brown. In ruderal habitats; local. ***Spergularia heldreichii*** is similar but with heart-shaped stipules, pink or lilac petals and smooth, **black seeds**. Coastal sands.

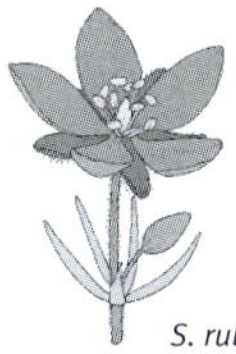

S. rubra

Spergularia rubra **SAND SPURREY** A **sticky-hairy annual or perennial** with spreading stems. Leaves ending in an abrupt point, not fleshy, **strongly whorled.** Stipules silvery and lanceolate. Petals more or less equalling the sepals, uniformly pink, to 6 mm. On sandy (not saline) soils. Local.

S. purpurea

S. fimbriata

Spergularia purpurea PURPLE SPURREY A short, slender annual or biennial to 25 cm high with slender stems, often growing in large masses. Leaves not fleshy, Stipules silvery and lanceolate. Flowers to 9 mm across, with **uniformly mauve petals that distinctly exceed the sepals**. Common in non-saline sandy, and waste places. ***Spergularia fimbriata*** is similar but more robust, woody below, with **long-awned leaves**, and **longer stipules to 12 mm**; on coastal sands. ***Spergularia diandra*** is similar to *S. purpurea* but **petals narrowly elliptical** (not oval) and lilac, bristly seeds; occurs in saline sandy and waste places.

S. rupicola

Spergularia rupicola A **robust perennial** with woody stems below, **glandular-hairy throughout**. Leaves narrowly linear, stipules ovate-triangular and rather silvery. Petals uniformly pink, more or less equalling the sepals. Seeds dark brown. Local, in maritime rocky places. ***Spergularia australis*** is similar but with leaves narrowed towards the base, **stipules very silvery** and black seeds. Habitats similar to those of *S. rupicola*.

S. inaperta

Silene inaperta A rather inconspicuous species with slender stems to 40 cm, erect, usually simple below and branched above, puberulent. **Leaves all narrowly linear**. Calyx to 10 mm with short teeth; **petals absent or within the calyx**. Local in grassy places.

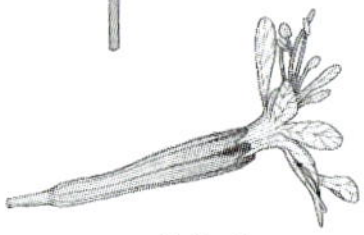
S. italica

Silene italica ITALIAN CATCHFLY An erect, branched perennial with a woody base, sticky above. Leaves ovate-elliptical, pointed and sticky-hairy, downy below. **Flowers erect**, to 20 mm across with deeply bilobed (often inrolled during the day), white petals, greenish or reddish below. Calyx 5–10 mm, purplish, cylindrical and hairy. Common in dry, grassy plac es. ***Silene patula*** is similar but with a more or less hairless calyx with dimorphic teeth, and reddish-purple petals.

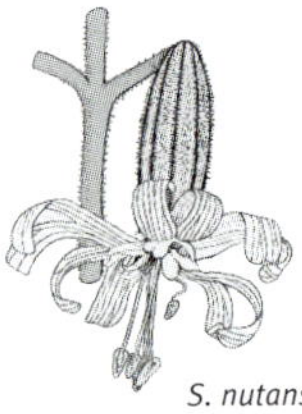
S. nutans

Silene rothmaleri An erect perennial with a woody stock, and lanceolate-spathulate leaves. Inflorescence lax, **flowers borne in panicles**, flowers whitish, the **calyx long to 23 mm**, and **stems sticky**. A rare endemic; only on sea cliffs in the Cape St. Vincent area.

Silene nutans NOTTINGHAM CATCHFLY A variable, erect perennial, usually unbranched and sticky above, downy below. Leaves spathulate at the base, to 10 cm, narrower and unstalked above. Flowers white, greenish or reddish below with deeply bifid petals and inward-rolling lobes in a loose, 1-sided and **drooping inflorescence**. Common in disturbed habitats.

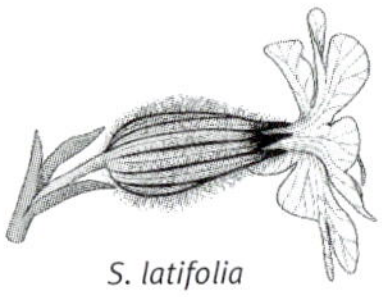
S. latifolia

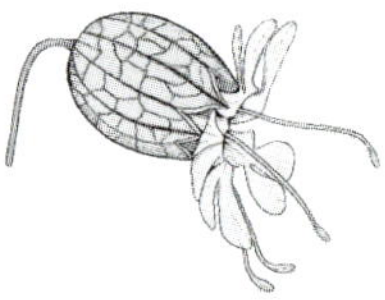
S. vulgaris

Silene latifolia WHITE CAMPION A densely **stickily hairy**, medium-tall perennial with ovate leaves, the lowermost stalked. Flowers large and white, to 30 mm across with deeply notched petals, the **calyx not inflated** and red-veined. Common on disturbed and cultivated ground, and on cliff tops. ***Silene vulgaris*** is similar but hairless or with downy stems, somewhat succulent leaves that are unstalked above, and an **inflated**, 20-veined bladder-like calyx with 6 teeth. Coastal environments and grassy habitats. ***Silene vulgaris* subsp. *maritima*** (= *S. uniflora*) has more diffuse stems with more widely spaced leaves. Coastal environments.

S. cretica

Silene cretica **CRETAN CATCHFLY** A slender, erect annual, hairy at the base, hairless and sticky above. Basal leaves ovate and broadest above the middle and stalked; upper leaves pointed and unstalked. Flowers pink, borne in lax, branched clusters with notched petals; **calyx hairless**. Common on cultivated and disturbed ground.

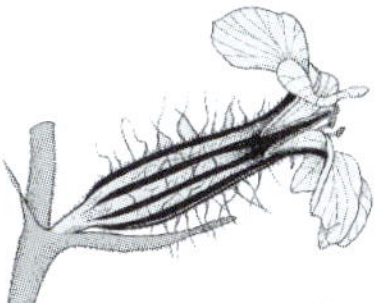

S. gallica

Silene gallica **SMALL-FLOWERED CATCHFLY** A downy annual with simple or branched stems to 45 cm, sticky above. Leaves ovate and stalked beneath, narrower and unstalked further up the stem. Flowers small, 6–12 mm across, white or pink, borne in a more or less **1-sided inflorescence**, the pedicels often exceeding the calyx. Calyx 7–10 mm, cylindrical to ovoid with teeth a quarter the length of the tube, sticky-hairy and 10-veined. Common in waste places throughout. ***Silene cerastoides*** is similar but the whole plant is **bristly hairy**, the pedicels not exceeding the calyx; calyx 8–10 mm long and strongly contracted above and below the capsule in fruit. Habitats similar to those of *S. gallica*.

1. *Spergularia purpurea*
2. *Silene rothmaleri*
3. *Silene latifolia*
4. *Silene vulgaris*
5. *Silene gallica*

Silene bellidiflora Similar to *S. gallica* but taller, to 60 cm; an erect, annual simple and bristly below, branched and pubescent above. Leaves sparsely bristly hairy. **Flowers more or less sessile** in a **dense, raceme-like inflorescence**. Flowers pink with deeply notched petals.

S. littorea

Silene littorea An ascending, much-branched, low to short, glandular-hairy, **sticky annual** with oblong to linear leaves, to 15 cm. Flowers with bright pink, notched petals, borne in a few-flowered, raceme-like, leafy inflorescence; calyx to 20 mm long, straight, or scarcely inflated above the middle when in flower. Very common on maritime sands. ***Silene psammitis*** is similar but has linear to **linear-lanceolate leaves** (not ovate). Habitats similar to those of the previous species, but this plant is much less frequent.

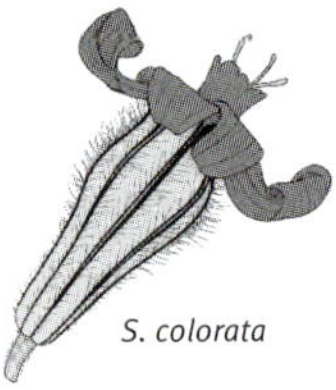
S. colorata

Silene colorata A variable, erect or spreading short to medium annual. Leaves ovate-linear, the lower leaves stalked. Flowers with bright pink (rarely white) deeply notched petals, 12–18 mm across, borne in lax clusters; calyx club-shaped, to 15 mm long. Very common or abundant on waste land and in sandy places throughout.

1

2

3

4

5

1. *Silene littorea*
2. *Silene colorata*
3. *Silene nicaeensis*
4. *Arthrocnemum perenne*
5. *Atriplex halimus*

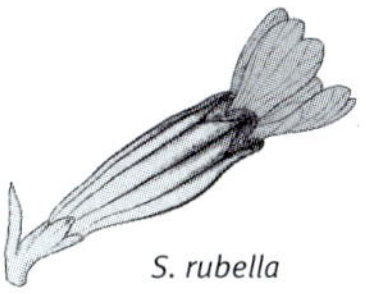

S. rubella

Silene rubella Stems 10–50 cm, lower leaves spathulate and pubescent. **Calyx whitish and red-tinged**, with ovate and blunt teeth. Petals bright pink. On arable and cultivated land.

S. nicaeensis

Silene nicaeensis A short to medium annual, **hairy throughout**, with stout, branched and sticky stems. Lower leaves narrowly spathulate, upper leaves linear-lanceolate, often curved. Flowers borne on pedicels equalling or shorter than the shortly hairy calyx, which is greenish-white with darker striping; **petals white** or faintly reddish-pink and deeply notched, erect, borne in clusters of up to 8 on a simple or branched inflorescence. Local on coastal sands in the south of the Algarve. Common on dunes to the west of Praia da Rocha. ***Silene obtusiflora*** is similar to *S. nicaeensis* but with thick, spatula-shaped leaves with hairy margins, and flowers borne in **branched raceme-like, flat-topped clusters**. ***Silene scabriflora*** is similar to *Silene obtusiflora* but the whole plant is **densely woolly hairy**.

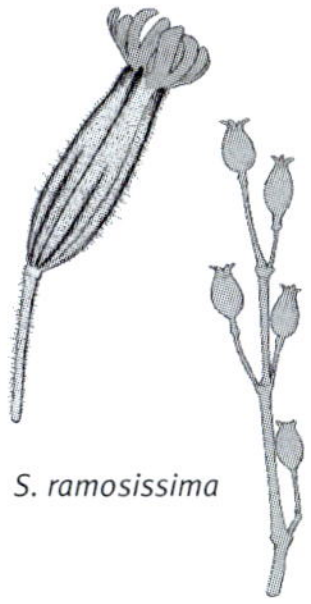

S. ramosissima

Silene ramosissima The whole plant **extremely sticky**, stems to 40 cm and very hairy. Leaves oblong-spatula-shaped, and pubescent. Inflorescence a long and slender, symmetrical, more or less flat-topped cluster, with a single pink flower on each axis, the lower pedicels to 5-times longer than the calyx. Local on maritime sands.

Kohlrauschia velutina KOHLRAUSCHIA Rather like a *Silene*, a short, glandular-hairy annual or biennial with pink, red-veined flowers with lobed petals borne in heads **surrounded by pointed, dry bracts**. On dry rocky slopes.

AMARANTHACEAE

As the result of recent phylogenetic analyses, this family now includes genera traditionally included in the family Chenopodiaceae. Herbs and shrubs with alternate, often succulent leaves. Flowers greenish with a 3–5-lobed perianth, ovary superior. Mostly halophytic or ruderal, often in maritime habitats.

Arthrocnemum perenne A small, very succulent mat-forming shrub, usually less than 1 m tall with creeping, subterranean stems, green, becoming red or brownish. Leaves opposite, scale-like and fused in pairs to form segments. Flowers tiny, borne in threes, each with 2 stamens; flowers falling to leave 3 scars in the segment. Common in salt marshes, maritime sands and shingle. ***Arthrocnemum fruticosum*** is very similar but with **stems erect to 1 m,** stout and non-rooting, and **glaucous.** ***Arthrocnemum glaucum*** (= *A. macrostachya*) is similar to *A. fruticosum* but shorter, to 50 cm tall, and becoming yellowish or reddish.

A. halimus

Atriplex halimus SHRUBBY ORACHE An erect, woody shrub to 2 m. Leaves **silvery-white,** small, to 30 mm long, alternate, narrowly ovate, and more or less unlobed. Flowers yellowish, borne in leafless branched terminal spikes. Common or abundant on sandy shores and estuaries throughout.

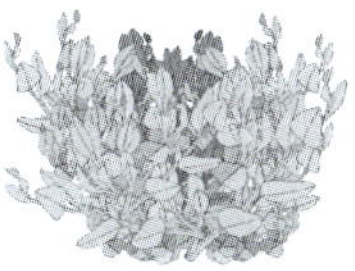

A. halimus

Atriplex prostrata SPEAR-LEAVED ORACHE A variable, tall, hairless and branched annual, mealy when young. **Leaves with spreading triangular basal lobes,** green-grey. Common on sandy shores.

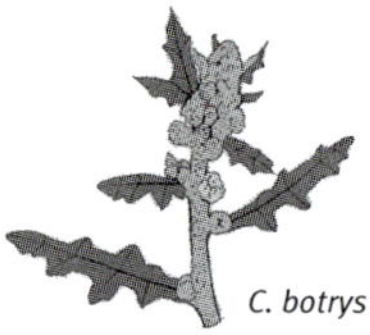
C. botrys

Chenopodium botrys STICKY GOOSEFOOT A short, **sticky** glandular-hairy aromatic annual with alternate, oblong, pinnately divided leaves. Flowers small and green, in long, narrow clusters. Common on cultivated and waste ground. ***Chenopodium murale*** is similar but taller, not hairy and with diamond-shaped leaves, in habitats similar to those of *C. botrys*. ***Chenopodium album*** is similar to both species, and is a deep green annual with reddish stems and a grey, dust-like covering.

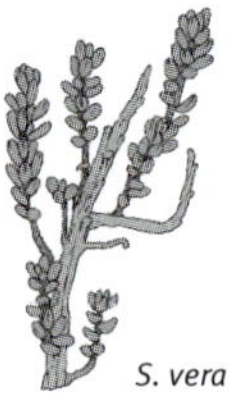
S. vera

Beta vulgaris SEA BEET A variable, hairless, reddish and rather fleshy annual or perennial to 1 m, but often prostrate. Leaves oval to lanceolate, dark green, leathery and shiny, and more or less untoothed, to 20 cm. Flowers small and green or purplish, borne in dense, leafy clusters in a leafless spike; stigmas mostly 2. Common on maritime sands and shingle.

Salicornia europaea Similar to, and often confused with, *Arthrocnemum* spp., but a **short annual**. Stems translucent, ascending and green or glaucous or yellowish, flushed with red when in flower. Common in salt marshes and on maritime sands. The young stems are edible.

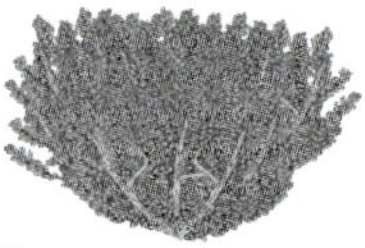
S. vera

Suaeda vera SHRUBBY SEABLITE A small succulent, woody shrub to 1.2 m, **leaves densely crowded, blunt and fleshy**, alternate, small, 5–18 mm, semi-cylindrical, sessile and glaucous, becoming reddish or purplish. Flowers small, greenish, borne 1–3 in the axils of the upper leaves. Common on sea cliffs and maritime slopes. ***Suaeda maritima*** is a similar **annual** with sparser, pointed, alternate leaves which are concave above in cross-section; occurs on maritime mudflats. ***Suaeda albescens*** is similar but has **leaves with a distinct whitish-grey bloom**. Common on the coast around Faro.

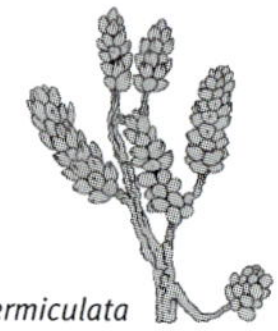
S. vermiculata

Salsola vermiculata A succulent **shrub** to 1 m with **very small, crowded leaves** to 5 mm long, semi-cylindrical, ovate at the base and semi-clasping the stem, **often pubescent**. Flowers small and green or pink-tinged, the perianth to 12 mm across in fruit, stigmas shorter than the style. Common on sea cliffs, maritime slopes and other coastal habitats. ***Salsola soda*** is similar, but a hairless **annual** with clasping, half-cylindrical, much **longer leaves to 70 mm**; occurs on sandy shores. ***Salsola kali*** is a similar succulent annual with **spine-tipped, linear-cylindrical leaves** and solitary whitish flowers; grows in a range of maritime environments.

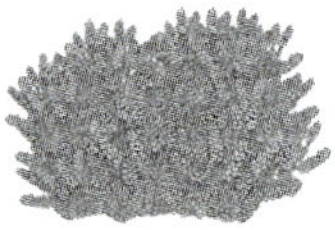
S. vermiculata

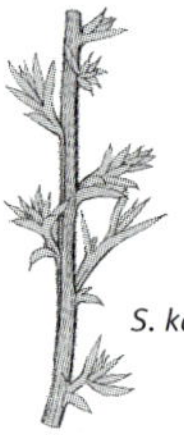
S. kali

Halimione portulacoides SEA PURSLANE Rather similar to *Atriplex halimus*; a low shrub with spreading and often rooting branches, more or less mat-forming. Leaves upward-pointing, silvery, mostly opposite, oblong and untoothed. Flowers small and green or reddish, borne in more or less leafless panicles. Common in salt marshes.

PORTULACACEAE

Annual or perennial herbs, often fleshy with cosexual flowers with only 2 opposite sepals and 3–6 petals. Fruit a capsule.

M. fontana

Montia fontana BLINKS A rather inconspicuous, tiny spreading or patch-forming annual with ovate, opposite leaves. Flowers white, tiny, borne in small loose clusters with stems that lengthen in fruit. Local on bare, damp ground.

CACTACEAE

A family of often spiny succulents (cacti) from the Americas, often with showy flowers. Cultivated and naturalised in the region. Now established to be distantly related to the pink family (Caryophyllaceae).

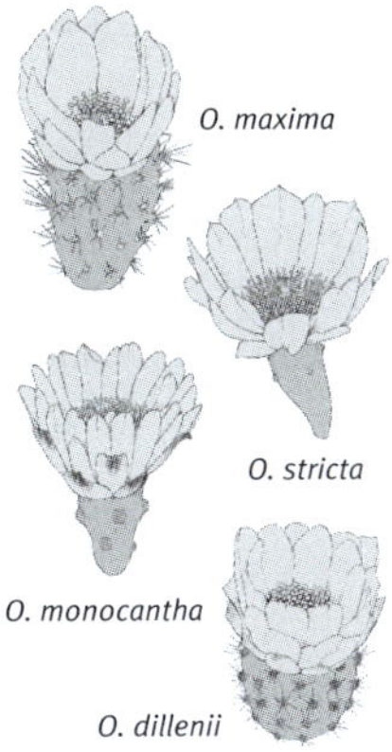

Opuntia maxima* (= *O. ficus-indica*)** PRICKLY PEAR A robust, blue-green, typically cactus-like plant **to 5 m**; woody and trunk-like at the base (rather tree-like) with **large, flattened jointed stem sections** to 50 cm long, and soon-falling inconspicuous leaves. Bristles hooked and yellowish, straight spines absent. Flowers bright yellow, to 1 m long. Fruit egg-shaped, to 90 mm long, yellow or reddish when ripe; edible. Common near towns. ***Opuntia stricta is similar to *O. maxima* but **smaller to 1 m**, the joints only to 15 cm, with yellow spines (not bristles). Common in disturbed and cultivated places. ***Opuntia monocantha*** is distinguished by its *bright*, **shiny green joints** and red-purple fruits to 60 mm. Less common than the previous two species. ***Opuntia dillenii* (= *O. tuna*)** has readily detachable joints to 1 m with conspicuous brown cushion-like clusters (areoles). Fruit red and small, to 30 mm. Cultivated and perhaps naturalised in the region.

1. *Chenopodium album*
2. *Beta vulgaris*
3. *Suaeda vera*
4. *Opuntia maxima*

1

2

3

4

5

6

AIZOACEAE

Herbaceous plants with simple, opposite or alternate, succulent leaves. Flowers actinomorphic appearing singularly or in few-flowered cymes developing from the leaf axils. True petals are absent, but some species have numerous linear petals derived from staminodes. Fruit a capsule with one to numerous seeds.

Carpobrotus edulis **HOTTENTOT FIG** A low succulent, trailing, mat-forming perennial with woody stems. Leaves opposite, 3-angled and upwardly curving, finely toothed along the edge and tapered towards the apex. Flowers large, to 10 cm across, solitary, virtually stalkless, very showy, bright pink or yellow with yellow stamens. Fleshy, edible fruit. Native of South Africa, very commonly naturalised and highly invasive in coastal areas. ***Carpobrotus acinaciformis*** is very similar but with **leaves blue-green and broadest above the middle**; flowers magenta with **purple stamens**. ***Carpobrotus chiliensis*** possibly occurs in the area, with smaller flowers 20–50 mm across.

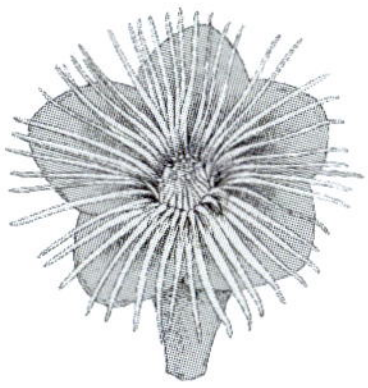

M. crystallinum

Mesembryanthemum crystallinum **ICE PLANT** A spreading annual covered in glistening, frost-like **crystalline hairs**. Leaves fleshy, oval, alternate and untoothed. Flowers solitary, 20–30 mm across, virtually stalkless, yellowish or whitish. In sandy, rocky and saline environments. ***Mesembryanthemum nodiflorum*** is similar but less crystalline and with more **cylindrical-linear leaves** that are red-tinged, and smaller flowers to 15 mm across. Habitats similar to those of *M. crystallinum*.

M. nodiflorum

Aptenia cordifolia is a mat-forming perennial with slightly fleshy, bright green, ovate leaves that are pointed at the apex and heart-shaped at the base. Flowers bright magenta and typically *Mesembryanthemum*-like. An ornamental, native to South Africa but widely planted and possibly naturalised in the Algarve.

Tetragonia tetragonioides A trailing or scrambling plant with triangular leaves to 50 mm long. Flowers small and yellow with 5 petals. Fruit horned. Native to New Zealand and Australia but cultivated as a source of 'spinach' and possibly naturalised in the area. Tolerant of saline environments.

NYCTAGINACEAE

Tropical or warm temperate trees, shrubs and woody climbers. Flowers with showy bracts.

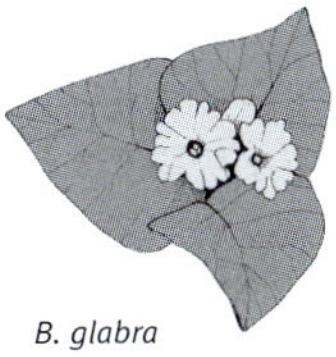

B. glabra

Bougainvillea glabra **BOUGAINVILLEA** A vigorous woody, virtually hairless climber to 10 m. Leaves opposite or alternate, to 60 mm long, untoothed and stalked. Flowers in threes, insignificant, whitish-yellow and funnel-shaped, surrounded by large, flower-like crimson bracts, 1 to each flower. Widely planted but rarely naturalised. ***Bougainvillea spectabilis*** is very similar but with leaves that have a densely hairy underside, and flowers with straight hairs and a purplish, not greenish, outer corolla. Bracts red, orange or violet.

1. *Carpobrotus edulis* pink form
2. *Carpobrotus edulis* yellow form
3. *Carpobrotus edulis*
4. *Carpobrotus edulis*
5. *Aptenia cordifolia*
6. *Bougainvillea glabra*

Order: SANTALALES

SANTALACEAE

A small family of woody or herbaceous hemiparasites that grow on the roots of surrounding vegetation. Flowers small, cosexual or unisexual.

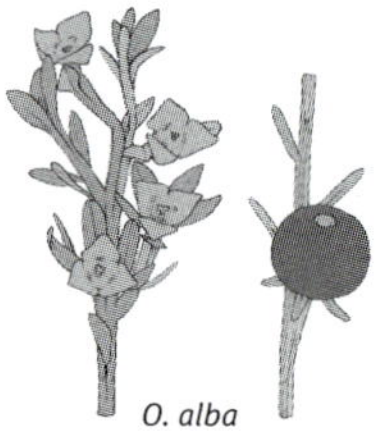

O. alba

Osyris alba A superficially broom-like, small yellowish shrub to 1 m. **Leaves small**, to 20 mm long, alternate, linear and entire, with a single mid-vein. **Bracts green, leaf-like and persistent in fruit.** Flowers sweetly scented, yellow and 3-parted; the male in small clusters, the female solitary and borne on separate shrubs. Fruit a red berry. Local on scrub, probably often confused with *O. quadripartita*.

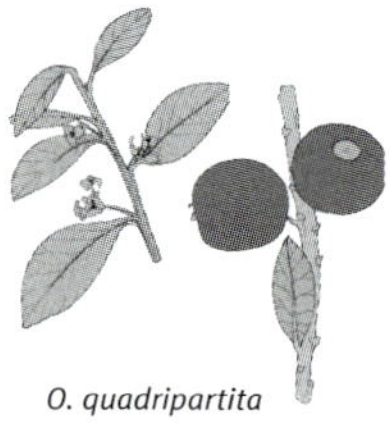

O. quadripartita

Osyris quadripartita Very similar to *O. alba* but taller, to 2.5 m, **leaves larger** (to 40 mm) and yellowish, leathery with pronounced pinnate veins (not a reliable diagnostic in all populations) and **small, papery bracts that are shed before fruiting**. **Flowers 3–4-parted** and androdioecious (male and cosexual flowers borne on separate plants). The most common species in the region; frequent on coastal *matos*. Populations in the Algarve sometimes exhibit unusually narrow leaves with indistinct pinnate leaf venation similar to *O. alba*, in which flower morphology and bracts subtending fruits are a more reliable diagnostic.

Order: SAXIFRAGALES

PAEONIACEAE

Herbs with alternate, divided leaves. Flowers solitary, large and showy, with usually 5 sepals and 5–13 petals, stamens numerous. Fruit consisting of 2–5 fleshy follicles, with large black or brown seeds.

P. broteroi

Paeonia broteroi A medium, **almost hairless perennial** with untoothed lower leaves with 17 or more hairless, narrowly elliptical leaflets that are grey-green below. Flowers solitary and large, to 10 cm across, purplish-red, with numerous yellow stamens. Follicles softly hairy. Local on rocky, forested hill slopes.

P. officinalis subsp. *humilis*

Paeonia officinalis* subsp. *humilis Superficially similar to *P. broteroi* but **leaves bluish-green and hairy beneath** and hairless above, and with a deeply channelled, **pubescent petiole.** Flowers red with **red filaments.** In meadows and bushy places.

SAXIFRAGACEAE

Herbs, usually perennial with alternate leaves in spirals or as basal rosette. Flowers usually cosexual and actinomorphic in cymes, racemes or solitary. Sepals and petals usually 4–5; stamens 3–10; ovary usually superior with 2–5 carpels. Friut a capsule or follicle.

Saxifraga granulata A perennial with most leaves in a basal rosette; leaves stalked, kidney-shaped, to 40 mm long and hairy. Flowers white, to 30 mm across, borne singly or in a loose inflorescence on erect, slender stems to 60 cm tall. Locally common in the Monchique area, and on the track to Picota.

CRASSULACEAE

Succulent annuals and perennials with leaves alternate, opposite or whorled. Flowers actinomorphic, star- or bell-shaped with up to 20 petals and an equal number of stamens, or twice as many. Fruit a cluster of carpels.

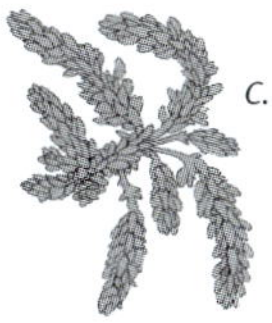

C. tillaea

Crassula tillaea A minute, moss-like plant with prostrate stems. Leaves to 2 mm, ovate and crowded. Flowers borne in small groups in the leaf axis, with petals 1 mm and shorter than the sepals; white or pale pink. Common in seasonally damp places.

1

3

2

4

5

1. *Osyris quardripartita*
2. *Osyris quardripartita*
3. *Osyris quardripartita*
4. *Paeonia broteroi*
5. *Saxifraga granulata*

1. *Sedum sediforme*
2. *Sedum sediforme*
3. *Sedum forsterianum*
4. *Sedum album*
5. *Sedum andegavense*
6. *Umbilicus rupestris*
7. *Cynomorium coccineum*
8. *Cynomorium coccineum*

Sedum sediforme* (= *Sedum nicaeense*)** A very succulent, short to medium perennial with flowering and non-flowering shoots, woody at the base, and reddish throughout. Leaves grey-green, later red and borne in spiralled rows. Flowers yellowish with 5–8 petals, borne on tall stalks with a dense sub-spherical inflorescence, then with recurved branches when mature, concave in fruit. Very common in dry rocky or sandy habitats throughout. ***Sedum tenuifolium is similar but with small grey-green leaves closely overlapping and clasping the stem at the base. ***Sedum forsterianum*** is similar to *S. sediforme* but has denser, slenderer blue-grey **leaves which are flat on top**, forming distinctive terminal clusters on non-flowering shoots. Common at higher altitudes, for example in the Monchique area, and on Picota.

Sedum brevifolium A low, fleshy, grey-pink to red perennial with **hairless, globose leaves.** Flowers pinkish-white, to 5 mm across with 5–6 petals borne in branched inflorescences.

Sedum album **WHITE STONECROP** A low to short, hairless, tufted and **mat-forming perennial** with creeping stems. Leaves alternate, linear-cylindrical to egg-shaped, to 12 mm long, **rather shiny and (usually) reddish**. **Flowers white**, borne in flat-topped clusters; to 9 mm across with 5 petals and 10 stamens. Rocky habitats; common on Picota and elsewhere in the Monchique area. ***Sedum andegavense*** is reddish, low and also has white or pinkish flowers but with **short stalks** and the plant glandular-hairy, at least below.

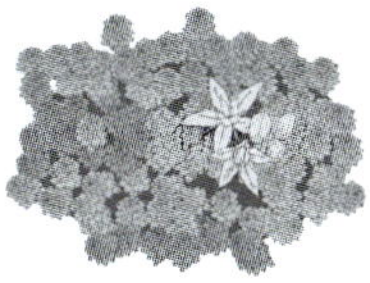

S. acre

Sedum acre **WALLPEPPER** A low, succulent, tufted, bright yellow-green hairless perennial. Leaves small, to 3 mm, ovate, blunt and overlapping. **Flowers bright yellow**, borne in small clusters. In rocky and sandy habitats.

Umbilicus rupestris **NAVELWORT** A fleshy, hairless perennial with distinctive basal circular leaves with a central hollow, borne on long stalks; upper leaves smaller, more kidney-shaped and with rounded teeth. Flowers whitish-green or yellowish, sometimes pink, tubular, and **pendent**, borne in long, tapered racemes exceeding their stems. Common on rocks, cliffs and as an epiphyte on planted palm tree trunks.

CYNOMORIACEAE

Previously confused with the Balanophoraceae; a poorly understood family of holoparasites growing on the roots of various Amaranthaceae. Fruit a small, single-seeded nut. Flowers unisexual or hermaphrodite, borne in a dense, highly reduced brush-like inflorescence.

C. coccineum

Cynomorium coccineum **MALTESE FUNGUS** A highly distinctive blackish-red plant sprouting as a club-shaped structure from an extensive underground rhizome system with lanceolate scale leaves. Inflorescence cylindrical with very dense, tiny flowers. Parasitic on halophytes in saline areas such as salt marshes and sea cliffs. Very local and much-declined in the region as a result of tourism-driven development; small populations occur on unstable cliff tops to the west of Praia da Rocha and possibly in the salt marshes of Alvor, but much declined there.

Order: VITALES

VITACEAE

A large, primarily tropical family of climbers with tendrils opposite the leaves. Flowers small and 4–5-parted, borne in clusters; calyx small, petals present. Fruit a berry.

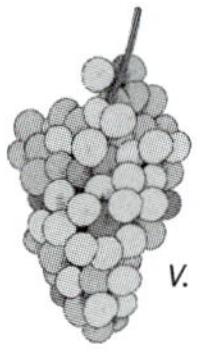

V. vinifera

Vitis vinifera GRAPE A climbing shrub with alternate leaves and tendrils opposite; tendrils branched. Leaves long-stalked and palmately 5–7-lobed, coarsely toothed. Flowers small, greenish, cosexual, borne in clusters hanging when mature. Fruit a berry (grape). Cultivated throughout the region, or naturalised as a relic of former cultivation.

Order: GERANIALES

GERANIACEAE

Herbs with alternate palmately lobed or pinnately lobed leaves. Flowers in cymes, umbels or solitary, 5-parted, usually more or less actinomorphic. Fruit with 5 1-seeded portions united into a prominent beak.

G. pyrenaicum

Geranium pyrenaicum A short, erect, hairy perennial with palmately lobed leaves to 50 mm across, divided over more than half the radius into 5–7 lobes with straight, entire margins, wavy towards the tip. Flowers pale purple or lilac with **deeply notched petals**. Local in dry, grassy habitats.

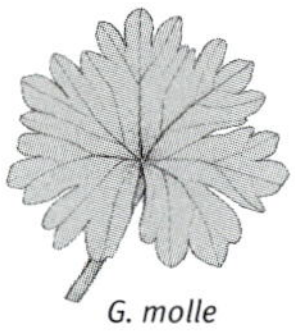

G. molle

Geranium molle DOVE'S-FOOT CRANE'S-BILL A low to short, sprawling annual with stems branched from the base, **grey-green and softly hairy**. Basal leaves long-stalked, rounded or kidney-shaped, divided over 70% of the radius into 5–7 wedge-shaped, 3-lobed segments; upper leaves more deeply divided and short-stalked or unstalked. Flowers pink-purple, to 10 mm across, borne in lax clusters, the **petals deeply notched**. Fields, sand dunes and cultivated land. ***Geranium rotundifolium*** is similar to *G. molle* but with more shallowly lobed leaves and bright pink flowers with **unnotched or slightly notched petals**. Habitats similar to those of *G. molle*.

G. rotundifolium

Geranium dissectum CUT-LEAVED CRANE'S-BILL A short to medium, spreading, hairy annual with ascending flowering stems. Leaves rounded in outline but **deeply dissected into 5–7 lobes** almost to the base. Flowers bright pink, to 10 mm across with shallowly notched petals; flower stalks less than 15 mm long; sepals spreading and with pointed tips. Fruit ridged and hairy. Common in a range of habitats, especially damp, disturbed, grassy places. Locally common throughout the region.

G. robertianum

Geranium robertianum HERB ROBERT A short to medium, hairy, **very aromatic** annual or biennial; usually **strongly flushed with red or purple**. Leaves palmate, the lower leaves with 5 pinnately lobed segments, the upper leaves with 3. Flowers to 18 mm across, pink or sometimes white, the petals slightly notched; pollen orange. Fruit hairy and ridged. Very common in cool, damp shady places and woods. ***Geranium purpureum*** is similar to *G. robertianum* but less flushed with red, flowers purplish-pink and smaller, to 14 mm across with yellow pollen, and petals usually unnotched. Habitats similar to those of *G. robertianum*.

G. purpureum

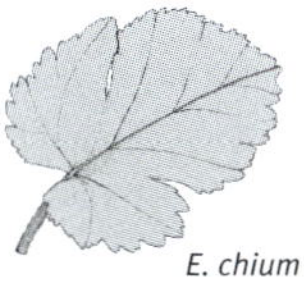

E. chium

Erodium chium A robust, low to medium, hairy perennial or biennial. Leaves ovate and divided into toothed, blunt lobes. Flowers pink-purple, 10–18 mm across, borne in 2–8-flowered clusters on non-glandular flower stalks; 2 petals slightly larger than the remaining 3. Fruit with short white hairs. Open, rocky places and roadsides.

E. botrys

Erodium botrys A short, hairy annual with an obvious stem above ground. Leaves bristly, to 50 mm across, ovate and deeply pinnately lobed and toothed, at least on the upper leaves. Flowers to 30 mm across, pink with darker veins, borne in clusters of up to 4; bracts brown. Fruit with a long beak, 4.5–11 cm. Dry rocky habitats and roadsides.

E. moschatum

Erodium moschatum **MUSK STORK'S-BILL** Rather similar to *E. cicutarium* but always stickily hairy and **smelling of musk**, leaflets only shallowly lobed (less than halfway to the midrib). **Stipules blunt.** Flowers larger, to 28 mm across, violet or purple. On cultivated and waste ground.

1. *Geranium dissectum*
2. *Geranium purpureum*
3. *Geranium purpureum*
4. *Erodium moschatum*

1

2

3

4

E. cicutarium

Erodium cicutarium **COMMON STORK'S-BILL** A very variable, low to medium, erect or prostrate, hairy (sometimes sticky), aromatic annual. **Leaves pinnately divided** without smaller lobes between the larger ones. **Stipules pointed** and whitish. Flowers purplish, pink or white, 7–18 mm across with up to 12 in a cluster, the upper 2 petals normally larger and with a blackish patch. Bracts brownish. Fruit hairy, with a **long beak** to 40 mm long. Very common in a range of open and disturbed habitats. **Subsp. *bipinnatum*** is a much smaller subspecies restricted to fixed, coastal sand dunes.

1. *Erodium cicutarium* subsp. *bipinnatum*
2. *Erodium cicutarium*
3. *Erodium malacoides*
4. *Lythrum salicaria*

E. acaule

Erodium acaule **STEMLESS STORK'S-BILL** Similar to *E. cicutarium* but *stemless* (though some populations of *E. cicutarium* appear stemless in exposed habitats). Leaves with petioles less than the length of the leaf-blade. Flowers lilac without darker patches, to 22 mm across. Fruit with white hairs and a beak to 50 mm long. In dry habitats, less frequent than *E. cicutarium*.

E. malacoides

E. laciniatum

Erodium malacoides **MALLOW-LEAVED STORK'S-BILL** An erect to sprawling short to medium, glandular-hairy biennial. Leaves oblong, heart-shaped at the base, toothed and sometimes 3- or more-lobed, covered in shiny glands. Flowers purplish-pink, 11–18 mm across, borne in 3–7-flowered clusters with at least 3 bracts at the base, borne on glandular-hairy stalks. Fruit beak to 35 mm long. Very common on bare or cultivated ground. ***Erodium laciniatum*** is similar to *E. malacoides* but with leaves indistinctly glandular-hairy beneath, and flowers in clusters of 4–9 with just 2 bracts at the base. Fruit beak longer, to 90 mm. On coastal sands.

Order: MYRTALES

LYTHRACEAE

Herbs with simple, alternate or opposite leaves. Flowers cosexual, actinomorphic, 4–6-parted and usually pink or purple; stamens 2–12.

L. junceum

L. hyssopifolia

Lythrum junceum A hairless, short to medium perennial with much-branched, sparse stems. Leaves mostly alternate, elliptical. Flowers purple, rarely white, solitary; petals 6, to 6 mm long; stamens 12, some or all protruding. Damp habitats. ***Lythrum hyssopifolia*** is similar but smaller and with 4–6 stamens, not protruding. Seasonally flooded areas and damp places.

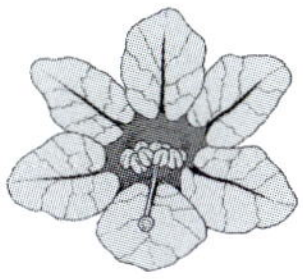

L. salicaria

Lythrum salicaria An erect, more or less hairless or shortly grey-hairy perennial to 1.5 m; stems sparingly branched with leaves in whorls of 3 or opposite; lanceolate. Flowers pink-purple, borne in long, dense terminal spikes; stamens 12, some or all protruding. Common in damp places and on riverbanks.

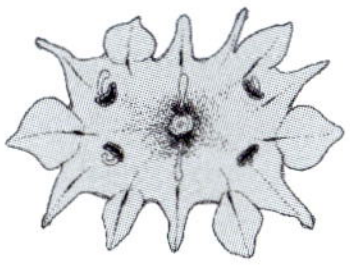

L. portula

Lythrum portula **WATER PURSLANE** A low to prostrate creeping, hairless annual with **stems rooting at the internodes**. Leaves opposite, fleshy, often reddish and tapered into a short stalk. Flowers purple, very small to 2 mm, solitary in the leaf bases; petals 6 or absent. Seasonally flooded and waterlogged muddy ground. ***Lythrum borystenicum*** is similar but **bristly**, at least when young, and with stems **not rooting at the internodes**.

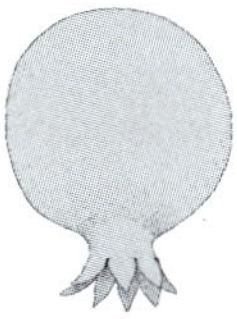

P. granatum

Punica granatum **POMEGRANATE** A deciduous shrub or small tree to 5 m with spiny, 4-angled young stems. Leaves opposite, shiny, bright green, roughly oblong, untoothed and virtually unstalked. Flowers to 40 mm across, with scarlet, crumpled petals and a fleshy calyx (hypanthium), borne in clusters of 1–3 near the ends of the branches. Fruit rounded, to 90 mm in diameter. Widely cultivated in the region, and naturalised.

ONAGRACEAE

Herbs with simple, alternate or opposite leaves. Flowers cosexual, actinomorphic, with 4 sepals and petals and a variable number of stamens. Fruit a capsule or berry splitting lengthways with distinctive cottony seeds.

E. hirsutum

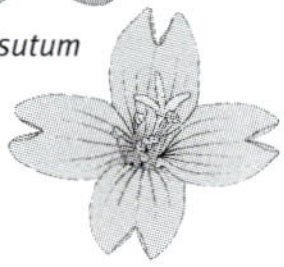

E. parviflorum

Epilobium hirsutum GREAT WILLOWHERB A robust, softly hairy perennial to 2 m. Leaves opposite, lanceolate, unstalked and partially clasping the stem; markedly toothed. Flowers to 25 mm with bright pink, notched petals, borne in a leafy raceme; stigma 4-lobed. Common in damp places and near rivers in the region. ***Epilobium parviflorum*** is similar but smaller, with leaves not clasping the stem and **small, pale pink flowers** to 12 mm. Local in damp places. ***Epilobium palustre*** has **untoothed leaves**, and pale pink to white flowers borne in lax, coarsely hairy racemes. Damp, wooded areas.

MYRTACEAE

A mainly tropical family of shrubs and trees with normally opposite, simple leaves. Flowers 4–5-parted with numerous stamens. Fruit usually a berry.

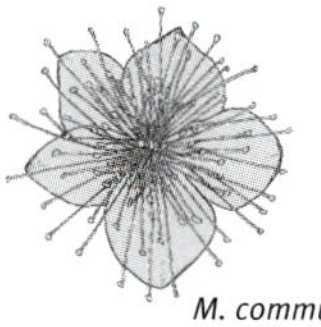

M. communis

Myrtus communis COMMON MYRTLE An erect, much-branched evergreen shrub to 2 m, glandular-hairy when young. Leaves opposite, shiny deep green, lanceolate and pointed, **aromatic when crushed**. Flowers white, to 30 mm with rounded petals and numerous, conspicuous protruding stamens. Berry blue-mauve then bluish-black when ripe. Common on the *matos*.

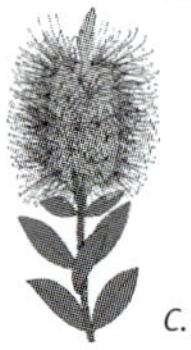

C. citrinus

Callistemon citrinus BOTTLEBRUSH An evergreen shrub to 2 m with stiff, arching stems. Leaves leathery green, narrowly elliptical, entire, and lemon-scented when crushed. Flowers crimson, borne in brush-like heads with prominent stamens. Native to Australia; widely planted.

Eucalyptus Evergreen trees native to Australasia with peeling bark and broad, erect leaves when juvenile; narrow and pendent when mature. Flowers borne in open clusters, with numerous, prominent stamens. Fruit a capsule. Widely planted and important for timber production in the region.

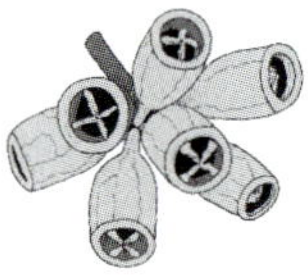

E. robustus

Eucalyptus robustus SWAMP MAHOGANY A large tree to 30 m with **persistent (not peeling) bark**. Juvenile leaves lanceolate, mature leaves narrower and tapered; shiny green. Flowers white, to 25 mm in clusters of 5–10. Fruit bell-shaped, to 12 mm across. Planted in swampy and slightly saline areas. ***Eucalyptus resinifera*** is taller to 40 m with **reddish bark** and egg-shaped fruits to 8 mm. ***Eucalyptus torquata*** is like *E. robustus* but a smaller tree, to 12 m and with **red, pink or orange flowers**. Planted as an ornamental.

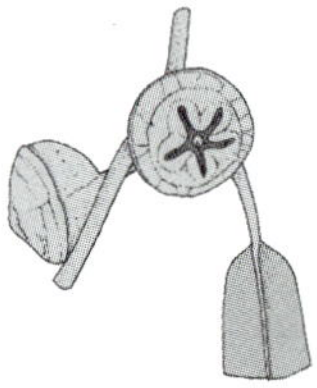

E. globulus

Eucalyptus globulus BLUE GUM A large tree to 40 m with smooth, peeling bark. Juvenile leaves ovate to lanceolate with a heart-shaped base, mature leaves narrower and tapered, to 30 cm long, grey, and aromatic when crushed. **Flowers solitary** and unstalked, white or pink, to 35 mm. **Fruit large** (to 30 mm across), wider than long; rounded and somewhat tapered towards the base. A native of Tasmania, very widely planted on hillsides in the region. ***Eucalyptus vimnalis*** is similar to *E. globulus*, with mature leaves 11–18 cm long, bark that peels in long ribbons, **flowers in umbels**, usually of 3 (not solitary), and a **small fruit** less than 10 mm across. Planted for timber and as an ornamental. ***Eucalyptus maidenii*** is like *E. vimnalis* but with 3–7-flowered umbels, **much larger leaves** to 20 cm long, and bluish fruits with 1–2 ribs.

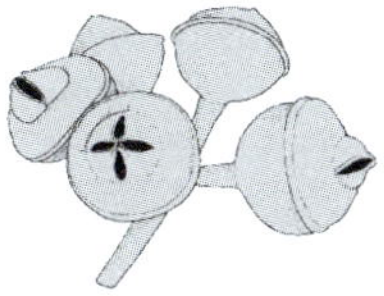

E. camaldulensis

Eucalyptus camaldulensis RIVER RED GUM A tree to 15 m with smooth, white, peeling bark. Juvenile leaves ovate to lanceolate, grey-green, mature leaves much narrower. Flowers yellowish-white, borne in clusters of 5–10. **Fruit hemispherical**, longer than wide, to 6 mm and with a broad, raised rim. Commonly planted for timber. ***Eucalyptus tereticornis*** is very similar to *E. camaldulensis* but produces larger, cone-shaped fruits to 10 mm.

Order: MALPIGHIALES

VIOLACEAE

A widespread family of herbs and shrubs poorly represented in the area. Flowers zygomorphic, solitary; sepals 5, separate; petals 5, the lowermost forming a lip, and extended behind into a spur. Stamens 5. Fruit a 3-parted capsule.

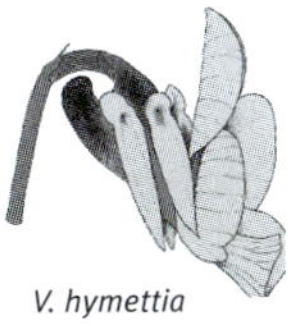

V. hymettia

V. arborescens

Viola hymettia A low, bristly hairy annual with tufted leaves that are ovate below, narrower above, and somewhat toothed. Stipules pinnately divided with a large terminal lobe. Flowers yellow with the upper petal flushed violet, or violet throughout. Grassy habitats; uncommon.

Viola arborescens TREE VIOLET A short, scrambling, shrubby, ascending perennial that is woody and often rather corky beneath. Leaves ovate to linear-lanceolate, pointed and slightly toothed. Stipules small and narrow, pinnately lobed. Flowers white flushed pale violet, to 15 mm, the spur short and blunt. Fruit erect and hairless when ripe. Local in shrubby thickets; roadsides on Cape St. Vincent.

1. *Epilobium hirsutum*
2. *Eucalyptus globulus*
3. *Viola arborescens*

EUPHORBIACEAE

A large family of herbs, shrubs, trees and lianas. Flowers unisexual on the same or different plants, inconspicuous and without petals.

Euphorbia **SPURGES** Monoecious; flowers in small groups surrounded by a cup-shaped involucre with 4–5 round or crescent-shaped glands, a single female flower and several 1-stamened male flowers, the whole structure forming a cyathium. Leaves normally alternate and entire. Plants exude a sticky, milky latex.

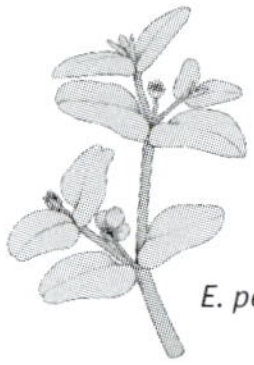

E. peplis

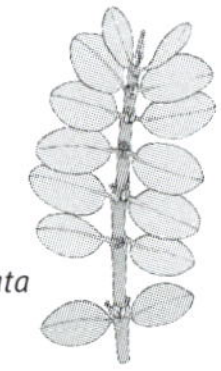

E. prostrata

E. maculata

Euphorbia peplis **PURPLE SPURGE** A flat, prostrate, **ground-hugging**, hairless annual with **4 main branches at the base**. Stems red or purple, leaves grey-green and small to 11 mm, opposite, oblong, and with a single rounded lobe at the base. Flowers tiny, greenish with semicircular red-brown glands, borne laterally or in clusters but not in umbels. Capsule nearly smooth and purplish. On sandy and shingly shores. ***Euphorbia prostrata*** is somewhat similar to *E. peplis* but with many branches at the base, stems that are hairy above, and capsules hairy on the keels. A North American weed naturalised locally in sandy places. ***Euphorbia maculata*** is similar to *E. prostrata* but with a **capsule entirely covered with closely appressed hairs**. A North American weed naturalised in ruderal places.

1. *Euphorbia exigua*
2. *Euphorbia paralias*
3. *Euphorbia paralias*
4. *Euphorbia peplus*

1

2

3

4

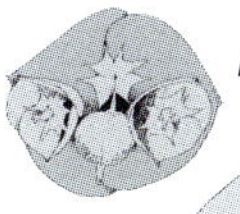
E. segetalis

E. pinea

Euphorbia segetalis A short, hairless **annual**, simple or with some branches at the base, to 30 cm. Leaves alternate and **narrow**, linear to linear-lanceolate. Rays 5, **bracts diamond-shaped** and yellow-green, glands with 2–4 horns. Capsule rough and glandular; seeds pale grey. In open, maritime habitats. ***Euphorbia pinea*** is very similar but a **perennial**, much-branched and woody at the base with densely leafy stems to 50 cm.

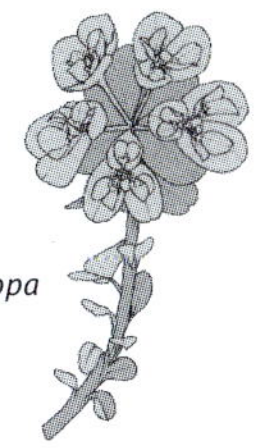
E. helioscopa

Euphorbia helioscopa **SUN SPURGE** A short, erect hairless annual, normally with a single stem, to 30 cm. Leaves ovate or spoon-shaped, broadest above the middle and **toothed** in the upper half. **Umbel 5-rayed, with 5 distinctive bracts** at the base that are yellowish and similar in shape to the leaves. Glands oval-shaped and untoothed. Capsule smooth and unwinged; seeds brown and wrinkled. A common weed on waste ground and bare soil throughout.

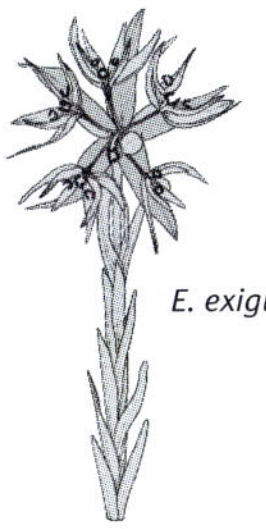
E. exigua

Euphorbia exigua **DWARF SPURGE** A very small, hairless, grey-green annual, normally to 10 cm, branched from the base. **Leaves very narrow**-lanceolate, untoothed and unstalked. Rays 3–5, up to 7-times branched, with narrowly triangular bracts. Glands crescent-shaped with 2 horns. Capsule shallowly grooved; seeds wrinkled and grey. Very common on cultivated, grassy and fallow habitats throughout.

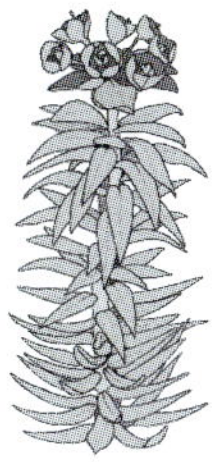
E. rigida

Euphorbia rigida **NARROW-LEAVED GLAUCOUS SPURGE** A hairless and distinctly bluish perennial with stout, erect to ascending stems arising from a **woody stock**, to 30 cm. Leaves fleshy, lanceolate and pointed, often flushed with purple. Rays 6–12, once or twice forked, ray leaves oblong. Glands **minutely horned. Capsule strongly 3-sided**; seeds smooth and whitish. Frequent on the *matos* and in waste places. ***Euphorbia broteri*** is very similar but with narrower linear-lanceolate leaves and **rough seeds**, on acid sands.

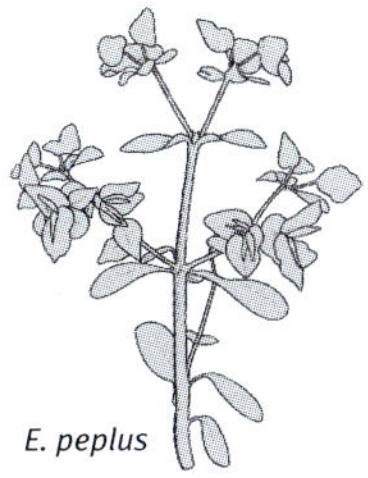
E. peplus

Euphorbia paralias **SEA SPURGE** A short to medium, hairless, stiffly erect, fleshy perennial forming clumps. Leaves grey-green, regularly and closely set on the stem and **overlapping**, ovate, **broadest towards the base** and concave above. Umbels with 3–6 rays, bracts ovate and concave. Glands kidney-shaped with long horns. Capsule rough along the back; **seeds smooth** and pale grey. Common in rocky and sandy maritime habitats and on dunes.

Euphorbia peplus **PETTY SPURGE** An erect small-medium, hairless annual to 30 cm, branched at the base. Leaves green, oval to rounded, untoothed and short-stalked. Umbels with **3 main rays**, with 3 bracts that are triangular-ovate to spoon-shaped, **green** and unstalked. Glands kidney-shaped with long, slender horns. Capsule smooth but with 2 ridges; seeds pale grey and pitted. Common on disturbed and waste ground. ***Euphorbia peploides*** may be merely a variant of *E. peplus* and is smaller with poorly developed umbels and scarcely pitted seeds. Dry habitats. ***Euphorbia falcata*** is similar to *E. peplus* but with unstalked leaves with a waxy bloom and an **unridged capsule** with **unpitted seed**. Habitats similar to those of *E. peplus*.

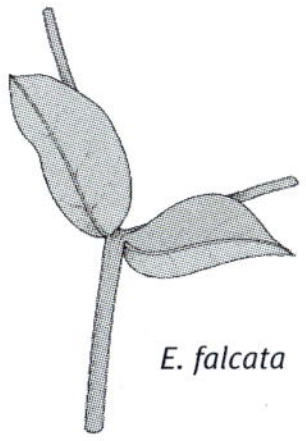
E. falcata

Euphorbia portlandica **PORTLAND SPURGE** A hairless perennial to 40 cm, rather similar to *E. paralias*, with stems prostrate to ascending (rather than stiffly erect). Leaves ovate, pointed and **broadest towards the tip**, tapered at the base, grey-green with a prominent midrib below. Umbels with 3–6 rays, bracts ovate below, triangular to diamond-shaped above. Glands yellow and kidney-shaped with prominent horns. Capsule deeply furrowed; **seeds pitted**. Sea cliffs and other maritime habitats.

Euphorbia transtagana A rather short, ascending, slightly glaucous and sometimes red-flushed annual to 20 cm tall (often less). Leaves more dense at the base, lax further up the stem; to 20 mm long, variably lanceolate to diamond-shaped and narrowed at base. Umbels with up to 5 rays; bracts broadly ovate to diamond-shaped, sometimes indistinctly toothed along the margin. Inflorescence (cyathium) short stalked; nectaries with 2 appendages. Seeds grey, avoid and ribbed. Rather uncommon; dry habitats.

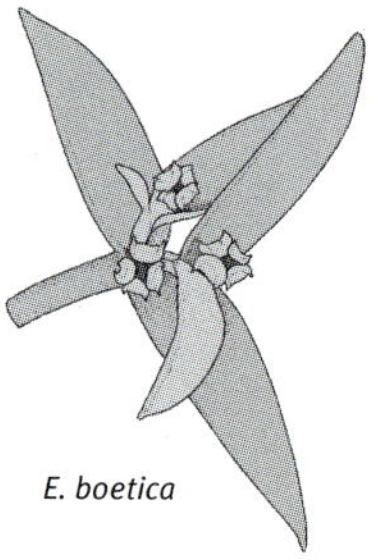
E. boetica

Euphorbia boetica A hairless, bluish or yellowish, tuft-forming perennial to 40 cm with a far-creeping rhizome. Leaves linear to linear-lanceolate, untoothed, and with 2–5 prominent veins beneath. Rays 4–6, branched up to 3 times. Glands variable in shape. Flower (cyathium) lobes **densely fringed with fine hairs**. Capsule scarcely grooved; seeds grey and pitted. Various habitats.

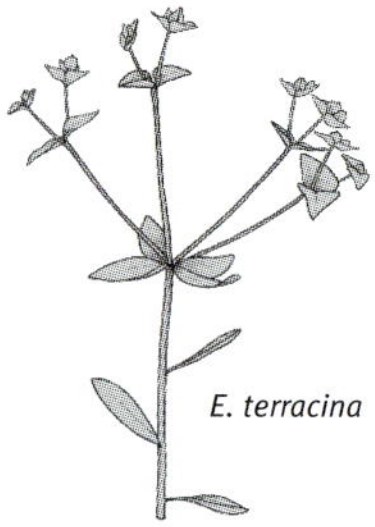
E. terracina

Euphorbia terracina A medium, hairless, **succulent perennial** with erect to ascending stems and non-flowering lateral branches. Leaves oblong to linear-lanceolate, minutely toothed, regularly and closely set on the stem and overlapping but **flat**. Umbels with 4–5 rays with as many oblong to diamond-shaped, green bracts. Glands with 2, long, slender horns. Capsule smooth; seeds pale grey and smooth with a boat-shaped, fleshy structure (caruncle) attached. Open coastal habitats.

Euphorbia serrata A short-medium greyish or bluish perennial with a woody stock. Leaves narrow-oblong with a finely **toothed margin**. Umbels with 3–5 rays, with as many lanceolate to rounded, **yellow bracts**. Glands oval-shaped, squared at one end. **Seeds pitted**. Dry habitats and *matos*.

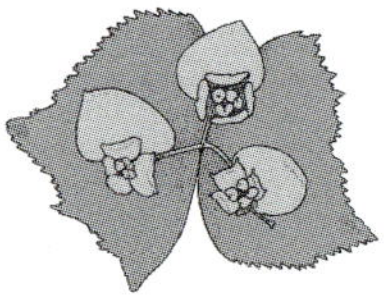
E. serrata

Euphorbia characias **LARGE MEDITERRANEAN SPURGE** A **large**, robust, imposing, **hairy** perennial with very thick, **unbranched stems**, to 1 m. Leaves bluish or greyish-green, linear to lanceolate, untoothed and crowded towards the upper part of the stems. Umbels with 10–20 short rays forming dense rounded or oblong heads at the apex of the stems. Bracts fused to form cups around the flowers, glands dark red-brown, notched or with short horns. Capsule smooth and softly hairy; seeds greyish. Native in exposed or shady dry places and also widely planted.

E. clementei

Euphorbia clementei A large, more or less hairless, tuberous perennial to 70 cm with cylindrical ascending stems that are woody at the base. Leaves oblong, **widest above the middle**, minutely toothed along the margins, **not pointed**, and narrowed into a very short petiole. Ray leaves ovate or diamond-shaped. Rays 5, **long** and outwardly spreading. Capsule hairless and grooved, with tubercles; seeds smooth, dark brown and shiny. Local on sea cliffs.

E. squamigera

Euphorbia squamigera A woody **shrub** to 1.2 m with much-branched stems always **bare below** and densely leafy above. **Leaves narrow**, linear-lanceolate, **pointed** and hairless or hairy beneath, entire or minutely toothed. Ray leaves rounded or diamond-shaped, also pointed. Rays 5. Capsule hairless and grooved, with tubercles; seeds smooth and brown. On sunny, rocky hillsides and sea cliffs.

Euphorbia welwitschii A large, robust, more or less hairless, tuberous perennial with stout stems to 60 cm, woody at the base. Leaves broadly ovate to lanceolate-ovate, **toothed** or more or less entire, to 30 mm. **Ray leaves broadly ovate**. Rays 5–6. Capsule covered in tubercles, sometimes sparsely hairy. **Seeds ovoid**, smooth and dark brown. Wooded and grassy slopes and field margins.

1. *Euphorbia portlandica*
2. *Euphorbia characias*
3. *Euphorbia characias*
4. *Euphorbia monchiquensis*
5. *Euphorbia monchiquensis*

1

2

3

4

5

Euphorbia monchiquensis A robust, tuberous perennial superficially similar to *E. welwitschii* but **larger**, to 1 m with longer, narrower leaves to 80 mm, lanceolate, **untoothed** and hairless or very sparsely hairy. **Ray leaves narrowly diamond-shaped**. Capsule sparsely covered in tubercles; seeds smooth and dark brown and slightly elongated. Local in cork oak forests and on roadsides near Monchique.

Euphorbia ingens CANDELABRA TREE A large, imposing, **cactus-like** perennial to 9 m tall. Stems bright green and 4-angled with irregular, spiny margins. Native to South Africa, occasionally planted in landscaped areas or cultivated in parks and gardens.

1. *Ricinus communis*
2. *Mercurialis annua*
3. *Linum bienne*
4. *Linum usitatissimum*
5. *Linum strictum*

1

2

3

4

5

Euphorbia pulcherrima **POINSETTIA** A shrub to 2 m tall with dark green leaves to 16 cm long and **leaf-like, showy red bracts**. A commonly planted ornamental native to Central America.

M. annua

Mercurialis annua **ANNUAL MERCURY** A dioecious, branched, erect annual to 50 cm, more or less hairless. Leaves to 50 mm long, opposite, ovate to elliptical, toothed and long-stalked, shiny green. Male flowers borne on dense, long, erect greenish spikes; female flowers few, borne in lateral clusters. Fruit to 3 mm, 2-lobed and bristly. A common weed on disturbed ground. ***Mercurialis tomentosa*** is densely woolly hairy, and superficially rather like a member of the Lamiaceae.

R. communis

Ricinus communis **CASTOR OIL PLANT** An imposing annual or shrub to 5 m, flushed red, bronze or purple. Leaves shiny, large and palmate, with 5–9 coarsely toothed lobes. Flowers in large terminal panicles with the male below with yellowish stamens, and the female above and with bright red stigmas. Fruit a 3-parted, spiny capsule, seeds bean-like. Planted in towns and occasionally naturalised.

LINACEAE

Small herbs with simple, untoothed leaves. Flowers 5-parted in a branched inflorescence, petals free. Stamens 5. Fruit a 10-parted capsule.

L. narbonense

Linum narbonense **BEAUTIFUL FLAX** A short to medium, hairless perennial with erect to ascending stems. Leaves alternate, greyish, linear and long-pointed, mostly 1- or 3-veined. **Bracts with membranous margins. Flowers large** with petals to 40 mm long, bright blue. Grassy places.

L. bienne

L. usitatissimum

Linum bienne **PALE FLAX** An annual to perennial herb with slender, erect to spreading stems, often **branched at the base**, to 60 cm. Leaves alternate, linear and long-pointed, mostly 3-veined. **Flowers pale blue or lilac**, borne in lax clusters, petals to 9 mm long, exceeding the ovate, hairy-margined sepals. Common in a range of habitats. ***Linum usitatissimum*** is similar to *L. bienne* but an **unbranched annual** with larger, **darker blue or white flowers** with petals to 18 mm long. Disturbed habitats.

L. maritimum

L. strictum

Linum maritimum A more or less hairy perennial with a woody stock and erect or ascending stems. Leaves lanceolate to narrowly elliptical, the **lowermost opposite**, 3-veined, the uppermost alternate and 1-veined. **Flowers yellow**, borne in lax clusters, the petals to 13 mm long. On damp, saline soils. ***Linum strictum*** is similar to *L. maritimum* but with narrowly lanceolate **leaves with inrolled margins** and flowers in a dense, spike-like inflorescence. Coastal sands and other dry places. ***Linum setaceum*** is similar to *L. strictum* but with linear leaves less than 0.5 mm wide, **leaves densely crowded** in the middle of the stem, and the inflorescence lax and much-branched.

Order: OXALIDALES

OXALIDACEAE

Herbs often with a bulbous stock, usually with 3-parted, clover-like leaves.

Oxalis pes-caprae BERMUDA BUTTERCUP A low, tufted perennial with numerous leaves arising from a bulbous stock; spreading and eventually carpet-forming; withering soon after flowering. Leaves **trifoliate** and clover-like. Flowers bright yellow and tubular with petals to 25 mm long. An abundant and highly invasive weed in coastal regions and disturbed land; increasing. *Oxalis articulata* is easily distinguished by its **pink flowers**. Local on disturbed and sandy ground; fairly common in hills around Monchique.

Order: FABALES

LEGUMINOSAE (also known as FABACEAE)

The third largest family of flowering plants, consisting of trees and herbs with trifoliate or pinnate leaves and pea-like, zygomorphic flowers with an upper petal (standard), 2 lateral wings which lie on the side of the 2 lower united petals (keel), concealing the 10 stamens and style. Fruit a dehiscent pod.

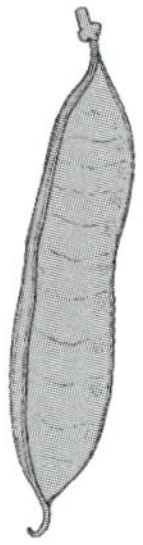

C. siliquastrum

Cercis siliquastrum JUDAS TREE A deciduous shrub or small tree to 10 m. Leaves heart-shaped to 12 cm long, blunt, long-stalked and hairless. Flowers pink (rarely white), to 20 mm long, borne in clusters arising directly from older branches, among previous blooms' pods, before the leaves appear, with a bell-shaped calyx. Pod pendent, linear-oblong, laterally flattened, to 10 cm, with a narrow wing along one edge. Widely planted as an ornamental.

Ceratonia siliqua CAROB TREE An androdioecious evergreen shrub or tree to 10 m with leaves pinnately divided into 2–5 pairs of dark green, rounded, leathery, untoothed leaflets; terminal leaflet absent. Flowers green or reddish and small with 5 sepals but without petals, **borne in lateral racemes directly from the trunk**. Pod large, to 20 cm, linear-oblong and laterally flattened, bluish-brown and pendent. Very common on the *matos* and field boundaries, a relic of cultivation throughout the region.

Albizia julibrissin ALBIZIA A deciduous tree to 15 m tall with a broad crown. Leaves large and twice pinnate with numerous oblong leaflets, hairy beneath. Flowers borne in large spherical heads of up to 50, long stalked; corolla tubular with 5 even teeth; greenish-white with a **conspicuous fringe of long, pink stamens** to 30 mm long. Pod oblong, to 40 mm long and with prominent seeds. Commonly planted as an ornamental.

A. dealbata

Acacia dealbata SILVER WATTLE A bushy tree to 30 m with smooth grey bark and silvery-hairy twigs and young leaves. **Leaves twice pinnate** with 8–10 pairs of primary divisions; leaflets 3–4 mm long, stipules rudimentary. Flowers pale yellow and fragrant, borne in heads 6 mm across that form large terminal panicles that exceed the leaves. Pod to 10 cm, linear-oblong and laterally flattened, bluish-brown. Native to Australia; widely planted as an ornamental and on roadsides. ***Acacia karroo*** is similar but with deciduous leaves and distinct stipules; native to South Africa but planted as an ornamental. ***Acacia mearnsii*** is similar to the previous species but shorter, to 15 m, with **yellow hairy young leaves**, and blackish pods distinctly constricted between the seeds.

Acacia retinodes A shrub or small tree to 10 m. 'Leaves' (phyllodes) lanceolate, wavy, light green and leathery with a single vein. Flowers pale yellow in heads to 5 mm across, borne up to 10 in a raceme. Pod to 12 cm, scarcely constricted between the seeds. Native to Australia, widely planted on roadsides throughout; common. ***Acacia cyanophylla*** is similar to *A. retinodes* but with **pendent twigs** and wider, **blue-green leaves**; pod distinctly constricted between the seeds. Planted to stabilise dunes and as an ornamental but less common than *A. retinodes*. ***Acacia pycnantha*** is similar to *A. cyanophylla* but with **sickle-shaped (not linear) leaves.** ***Acacia longifolia*** has flowers borne in **lateral spikes** and sub-cylindrical pods.

1. *Oxalis articulata*
2. *Oxalis pes-caprae*
3. *Cercis siliquastrum*
4. *Ceratonia siliqua*
5. *Ceratonia siliqua* PHOTO: FINN RASMUSSEN
6. *Acacia retinodes*

A. foetida

A. foetida

Anagyris foetida BEAN TREFOIL A deciduous, foetid shrub to 4 m. Leaves trifoliate with narrowly elliptical leaflets, silvery-hairy below and with papery stipules. Flowers yellow, borne in short clusters; calyx bell-shaped, bluish. Pod to 20 cm long. In dry, rocky places.

1

2

3

4

5

C. villosa

Calicotome villosa A superficially gorse-like erect shrub to 3 m with slender spines on the branches. Leaves trifoliate, silver-hairy below. Flowers yellow and large, to 18 mm long, borne in umbrella-shaped clusters. Calyx tubular with 5 teeth, the upper part falling in flower to leave a **cup-shaped structure**. Pod to 30 mm, hairy. Local on the *matos*.

Cytisus scoparius BROOM A familiar shrub without spines. Much-branched, to 2 m, stems long, slender and flexible, normally **5-angled and ridged**, hairy or not. Leaves small, trifoliate (single-foliate and sessile on young branches); ovate-elliptical, with **short petioles**. Flowers large, to 20 mm, golden yellow, solitary or paired. **Pod hairy along the margins only,** oblong, compressed, to 40 mm. Very common in wooded areas on acid soils. ***Cytisus baeticus*** is distinguished by its 7–8-angled, nodular stems, **petiolate leaves** and **pods covered in long, white, woolly hairs.** ***Cytisus grandiflorus*** is similar to *C. scoparius* but with **all leaves sessile**, the pod clothed in long, white, woolly hairs as in *C. baeticus*.

T. monspessulana

Teline monspessulana A shrub to 1 m, similar to *Cytisus* spp. but leafier. Leaves stalked and trifoliate with oval, variably hairy leaflets. Flowers deep yellow, borne in lateral clusters; calyx silvery-hairy and 2-lipped. Pod oblong, flattened, to 20 mm and white-woolly. Common on the *matos*.

Genista hirsuta* (= *Genista hirsuta* subsp. *algarbiensis*, = *Genista algarbiensis*)** A low, intricately spiny, gorse-like shrub to 50 cm with **unbranched spines on older stems**. Leaves simple, elliptical, with long hairs on the margins and lower surface, hairless above. Flowers yellow, borne in crowded terminal racemes. Very common on the *matos*, sometimes subdominant, forming swathes of yellow on hillsides when in bloom. ***Genista tournefortii is very similar but with **much-branched spines**. ***Genista scorpius*** has simple **leaves and flowers borne in clusters directly on the spines**. ***Genista triacanthos*** is a gorse-like shrub with robust stems bearing **tripartite spines** and clusters of sulphur-yellow flowers. Scattered in the region.

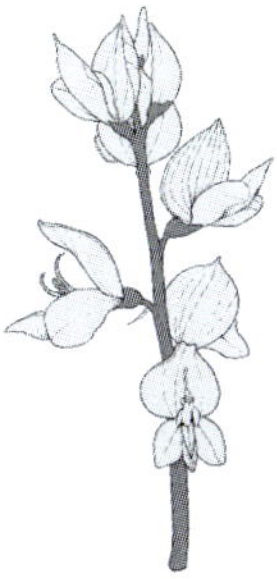

S. junceum

Chamaespartium tridentatum A dwarf, much-branched shrub to 50 cm with **distinctly winged young stems**; wings undulate and leathery, elongated at each node to form 3 small, tooth-like lobes. Flowers yellow, borne in crowded ball-like terminal clusters. Common on heaths, roadsides and on the *matos*.

Spartium junceum SPANISH BROOM A large, spineless, broom-like shrub to 3 m with cylindrical, blue-green, rush-like stems. Leaves sparse, linear-ovate and soon-falling. Flowers large, to 25 mm, and bright yellow, solitary but in large numbers; sweetly scented. Pod to 80 mm, flattened. Common on dry slopes and in woods; also planted.

---

1. *Cytisus scoparius*
2. *Genista hirsuta*
3. *Genista triacanthos*
4. *Chamaespartium tridentatum*
4. *Spartium junceum*

U. parviflorus

Ulex europaeus **GORSE** A stout, erect, very spiny shrub to 2 m, densely branched above and bare below. Young twigs and spines glaucous; twigs somewhat hairy. Leaves highly reduced, to 8 mm; spines conspicuous, straight and hairless. **Calyx 2-lipped** and persistent, yellowish; flowers large to 20 mm, yellow with straight wings longer than the keel. **Bracteoles at least 2 mm wide**. On acid heaths and in hedges. ***Ulex parviflorus*** is similar but with smaller bracteoles to 1.5 mm wide, and **flowers smaller (to 8 mm) not exceeding the calyx**. ***Ulex argenteus*** is similar but the **young shoots are covered in short white hairs**. There are 3 subspecies in the area: **subsp. *argenteus*** is low-growing and glaucous and occurs in *Quercus* woods; **subsp. *subsericeus*** is taller and not glaucous and occurs on sandy maritime soils; **subsp. *erinaceus*** is a low, compact, **silvery shrub** with a distinctly **hairy calyx** and occurs in the Cape St. Vincent area (this subspecies is sometimes treated as a true species, *Ulex erinaceus*).

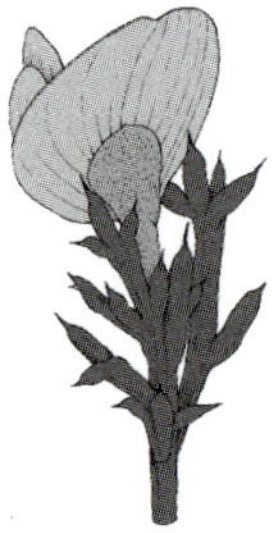

S. boivinii

Stauracanthus boivinii **STAURACANTHUS** Similar to *Ulex* but with mostly opposite leaves. A **very dense** spiny and intricately branched, cushion-like low shrub to 30 cm. Twigs hairless or with brown hairs, spines straight, to 4 mm, leaves tiny, to 3 mm and **spine-tipped**. Flowers yellow, the **petals much-exceeding the 2-lipped brown-hairy calyx. Pod small to 12 mm,** oblong, greatly exceeding the calyx. Rocky habitats and wind-swept scrub.

Stauracanthus genistoides Very similar to *S. boivinii* with flowers in which the **standard equals (not exceeds) the calyx**. Pod larger, to 25 mm. Dunes, rocky slopes and wind-swept *matos*. Two subspecies are recognised: **subsp. *genistoides*** has slender, flexible spines and a slightly hairy standard; **subsp. *spectabilis*** has stouter, rigid spines and a very hairy standard, and grows in the Cape St. Vincent area.

S. genistoides

Retama sphaerocephala **LYGOS** A much-branched spineless shrub to 2 m with erect to spreading branches that are **glabrous**, ribbed and have a **silvery sheen**. Leaves small and linear, silvery-hairy and soon falling. Flowers yellow, to 8 mm long, borne in lax, lateral clusters; calyx persistent in fruit. Pod smooth and egg-shaped, to 9 mm long. Dry slopes. ***Retama monosperma*** has **white flowers** to 12 mm long and **roughly hairy pods**. Common on roadsides.

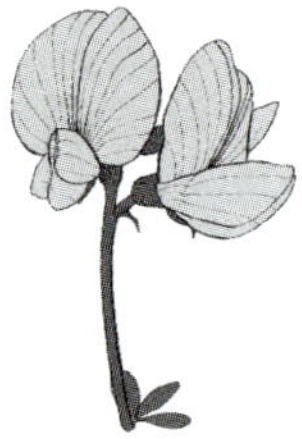

A. complicatus

Adenocarpus complicatus **ADENOCARPUS** An erect spineless, sparsely hairy shrub, rather broom-like but leafy. Leaves trifoliate with narrow elliptical, silkily hairy leaflets to 5 mm wide, **flowers orange-yellow**, to 15 mm in terminal clusters; calyx tubular, with a bilobed upper lip, and trilobed lower lip. Pod oblong-lanceolate. Common on scrub. ***Adenocarpus hispanicus*** is similar but with smaller leaflets less than 3 mm wide, with **inwardly rolled margins**. **Subsp. *argyrophyllous*** has leaves densely silvery-hairy above.

L. luteus

Lupinus luteus **YELLOW LUPIN** A robust, medium-tall hairy annual. Leaves palmately lobed with oblong leaflets. Flowers bright yellow, to 16 mm long borne in long whorls along the raceme; scented. Pod to 50 mm long, densely hairy, black when ripe. Common in sandy places, and cultivated for fodder.

L. micranthus

Lupinus micranthus **HAIRY LUPIN** A short, brown-hairy annual with palmate leaves of sparsely hairy oblong leaflets. Flowers blue, to 14 mm long, the standard with a whitish central region, and the keel tipped with black-violet; borne in irregular whorls along the raceme. Pod to 50 mm, **hairy** and red-brown when ripe. Common on cliff tops in the region. ***Lupinus albus*** is similar but with white to pale blue flowers and hairless pods that are yellow when ripe.

1. *Ulex europaeus*
2. *Stauracanthus boivinii*
3. *Stauracanthus genistoides*
4. *Lupinus angustifolius*
5. *Lupinus luteus*
6. *Lupinus micranthus*

L. angustifolius

L. varius

Lupinus angustifolius **NARROW-LEAVED LUPIN** Rather similar to *L. micranthus* but with **narrower leaflets** (less than 5 mm wide), linear to spatula-shaped, slightly hairy beneath. Flowers dark blue, borne alternately along the raceme. Pod with short hairs, yellow to black when ripe. Common in waste places and on cultivated ground. ***Lupinus varius*** is similar to the above species but with **larger flowers to 17 mm long**, the standard with a paler, white or yellow region, borne in irregular whorls along the raceme. Pod reddish-brown when ripe. Habitats similar to those of *L. angustifolius*.

1. *Astragalus boeticus*
2. *Astragalus boeticus* fruit
3. *Astragalus tragacantha* subsp. *vicentinus*
4. *Erophaca boetica* in fruit; inset *E. boetica* flowers

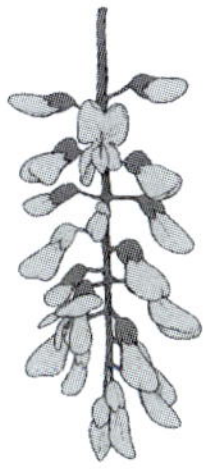

R. pseudacacia

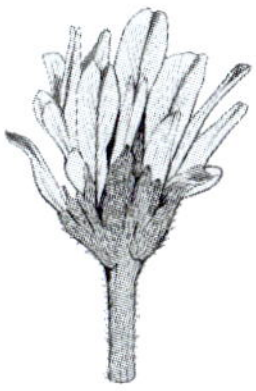

A. stella

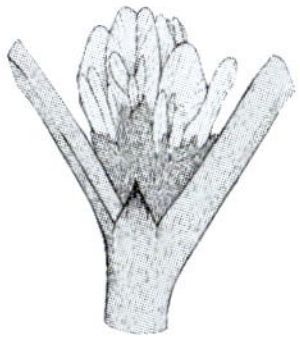

A. sesameus

A. echinatus

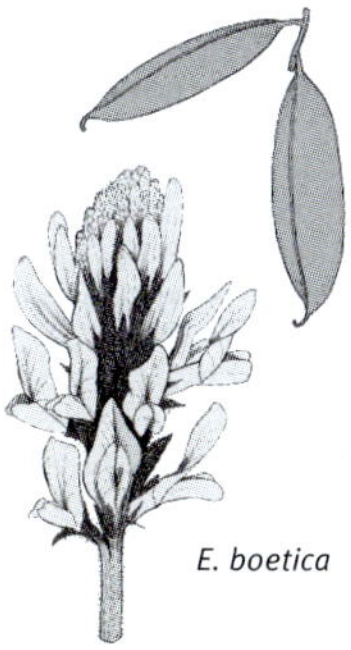

E. boetica

Robinia pseudacacia FALSE ACACIA A deciduous tree to 25 m with irregularly fissured, grey-brown bark, and spine-like stipules on younger branches. Leaves pinnate, more or less hairless. Flowers white, scented, with a standard yellowish at the base, borne in **pendent racemes**. Pod linear-ovate, to 10 cm, flattened. Commonly planted as an ornamental.

Astragalus boeticus A hairy, short-medium, erect annual with pinnate leaves of up to 15 pairs of ovate leaflets, notched at the tip, hairless above, and slightly hairy below. Flowers pale sulphur yellow and off-white, to 14 mm, borne on dense racemes on stalks as long as the leaves or slightly shorter; wings longer than keel. Pod oblong, triangular in section, shortly hairy, borne in dense clusters erect at first, then drooping. Common in sandy places and along cliff-top paths.

Astragalus stella A low to short annual to less than 20 cm with spreading or ascending stems. Leaves with 9–11 pairs of ovate leaflets, hairy on both surfaces. Flowers yellowish, to 11 mm long, borne on dense racemes half to the same length as the leaves. **Pods borne in star-shaped clusters,** more or less erect, almost straight, laterally compressed. Common on disturbed ground. **Astragalus sesameus** is similar but with leaflets only sparsely hairy, racemes scarcely stalked, pods more erect and flowers often bluish. Habitats similar to those of *A. stella*. **Astragalus epiglottis** is similar to *A. stella* but taller with less hairy leaflets, and racemes scarcely stalked.

Astragalus echinatus A short to medium somewhat hairy annual to 60 cm with leaflets in 6–9 pairs, notched at the tip, hairy beneath and hairless above. Purplish flowers borne in very dense racemes with stalks equalling or longer than the leaves. Pod ovate-triangular, flattened and covered in bristles, with a hooked beak.

Astragalus tragacantha subsp. vicentinus (= Astragalus massiliensis) An extremely spiny, intricately branched and cushion-like low shrub to 30 cm. Leaves with 6–12 pairs of elliptical leaflets ending in a spine, densely hairy beneath. Flowers white, borne in short racemes. Pod oblong, shortly hairy, to 10 mm. Locally common in exposed sunny and rocky places.

Astragalus algarbiensis A perennial with stems to 30 cm, erect or ascending. Leaves to 12 cm, leaflets 9–14 in pairs of **wedge-shaped leaflets**, hairy beneath, hairless above. Flower stalks more or less equalling the leaves; yellow flowers borne in dense racemes of pendent flowers. **Pod crescent-shaped, wrinkled, and hairless**.

Erophaca boetica (= Astragalus lusitanicus) An erect, robust perennial to 50 cm. Leaves grey-green with 8–12 elliptical leaflets, hairless above. Flowers cream or greenish, borne in lateral, long-stalked, dense and many-flowered racemes. Pod oblong, large and **inflated**, to 70 mm long and pendulous in clusters. Common on the *matos* and rocky places throughout the region.

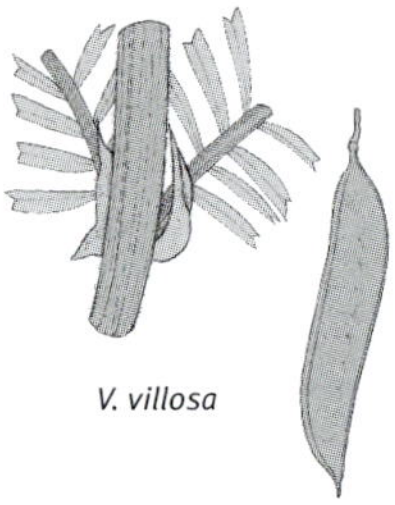

V. villosa

Psoralea bituminosa An erect, branched perennial to 10 cm, smelling of tar when crushed. Leaves long-stalked, trifoliate with narrow to broad untoothed leaflets to 60 mm long. Flowers blue-violet borne on long-stalked clover-like clusters. Pod ovoid, flattened with a sickle-like beak. Common on waste ground and disturbed habitats. ***Psoralea americana*** is similar but with **toothed leaflets** and flowers in longer racemes on stalks equalling the leaves, with white, violet-tipped flowers. Habitats similar to those of *P. bituminosa*.

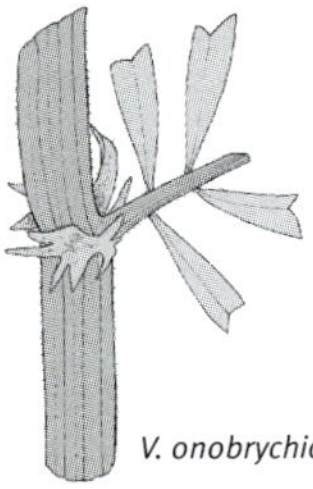

V. onobrychiodes

Vicia villosa **FODDER VETCH** A variable annual herb to 2 m, usually scrambling or climbing. **Leaves unlobed**, pinnate with 4–12 pairs of linear-elliptical leaflets, with hairy stipules and branched tendrils. Flowers violet to purple, often with creamy wings borne in conspicuous racemes of up to 30 flowers; stalk shorter than the subtending leaf. Pod brown and hairless. Common on cultivated and waste ground. ***Vicia benghalensis*** is very similar but smaller, often with lobed stipules, reddish-purple flowers that are **black at the tip**, with calyces convex at the base, borne in racemes longer than the leaves; pod at least partly hairy. Habitats similar to those of *V. villosa*.

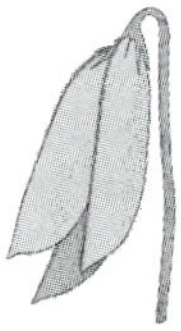

V. hirsuta

Vicia onobrychiodes **FALSE SAINFOIN** A hairy or hairless perennial to 120 cm. Leaflets in 4–11 pairs, linear to oblong and **narrow**, less than 4 mm wide. Stipules toothed. Flowers large, to 24 mm, borne in rather 1-sided, long-stalked racemes of 4–12 flowers, violet with a paler keel, to 25 mm. Pod more or less hairless. In disturbed and sandy waste places.

V. disperma

Vicia hirsuta **HAIRY TARE** A short to medium, delicate, hairy annual with leaves of 4–10 pairs of linear to oblong leaflets notched at the tip. Tendrils branched. Flowers white, tinged purple and small, to 4 mm long in 1–8-flowered racemes. Pod black, and rather hairy. Common on roadsides. ***Vicia disperma*** is similar but sparsely hairy with pale purple flowers and brown, hairless pods. ***Vicia tenuissima*** is similar but with 2–5-flowered racemes that exceed the leaves.

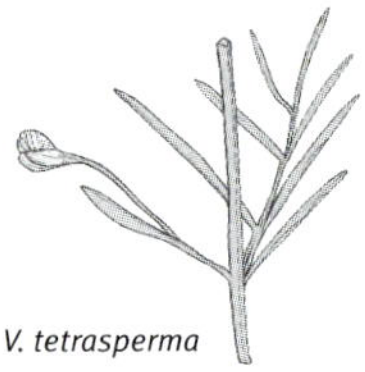

V. tetrasperma

Vicia tetrasperma A more or less hairless annual with up to 8 pairs of linear leaflets and entire stipules. **Racemes with 1 or 2 flowers**, about equalling (not exceeding) the leaves; corolla pale purple; calyx teeth unequal. Pod hairless. ***Vicia pubescens*** is similar to *V. tetrasperma* but slightly hairy, with up to 6-flowered racemes and hairy pods.

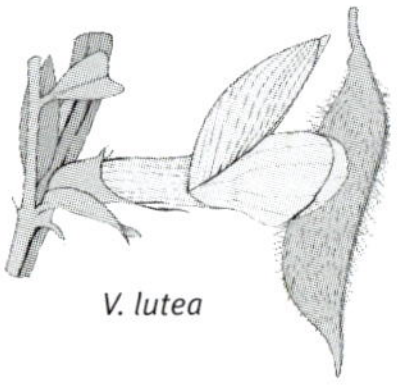

V. lutea

Vicia lutea **YELLOW VETCH** Superficially similar to the other *Vicia* spp. but distinguished by its **yellowish-white flowers**, sometimes flushed with purple with a **hairless standard**, and normally densely hairy pods. In coastal, sandy habitats. ***Vicia hybrida*** is very similar, but with **solitary flowers** with a dorsally hairy standard, borne on short stalks, and unequal calyx teeth. Damp, shady places.

1. *Psoralea bituminosa*
2. *Vicia benghalensis*
3. *Vicia lutea*
4. *Vicia sativa*
5. *Lathyrus cicera*

Vicia sativa **COMMON VETCH** A variable sprawling, hairy annual. Leaves with 3–8 pairs of linear to heart-shaped leaflets, **tendrils branched**, stipules toothed. Flowers purple, solitary or paired, to 30 mm long. Pod yellowish or blackish, dehiscent. The most common *Vicia* species; frequent in all grassy and disturbed habitats.

Vicia faba **BROADBEAN** A very commonly cultivated crop in the Algarve with grey-green foliage, white and black flowers and large, pendent pods.

Lathyrus cicera A slender, hairless annual to 80 cm with narrowly winged stems. Leaves with 1 (rarely 2) pair(s) of lanceolate-ovate leaflets, with simple or branched tendrils on the upper leaves. Flowers solitary on stalks to 30 mm, red, orange or brownish; calyx teeth equal, pod hairless, to 40 mm, and with 2 keels on the upper edge. Very common on the *matos* and disturbed habitats.

1

2

3

4

5

1. *Lathyrus ochrus*
2. *Ononis reclinata*
3. *Ononis reclinata*
4. *Ononis dentata*
5. *Ononis natrix*
6. *Ononis natrix*

L. tingitanus

Lathyrus ochrus A medium to tall hairless annual to 1 m with winged stems, and leaves borne on leaf-like stalks; lower leaves ovate-oblong, upper leaves with 1–2 pairs of leaflets. **Flowers pale yellow.** Pod to 60 mm with **2 wings** along the upper edge. Very common on the *matos* and disturbed habitats. **Lathyrus clymenum** is similar to *L. ochrus* but with narrower, **linear-elliptical leaves** to 11 mm long, and crimson or purplish flowers (rarely pale yellow), and a **pod grooved** on the upper side. Roadsides. **Lathyrus articulatus** is very similar to *L. clymenum* but with very narrow leaflets, usually less than 5 mm wide, and flowers with white or pink wings. **Pod not grooved** on the upper side.

Lathyrus tingitanus TANGIER PEA An annual herb with winged stems to 1 m. Leaves **with 1 pair of linear-lanceolate leaflets**; tendrils branched. Stipules arrow-shaped at the base; ovate. **Flowers bright red-purple** in racemes of 1–3. Common on the *matos*.

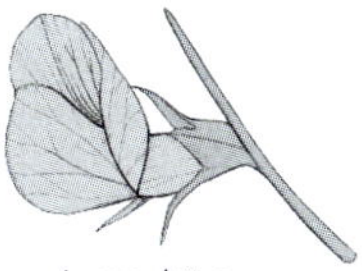
L. angulatus

Lathyrus angulatus A short to medium hairless or hairy annual similar to other *Lathyrus* spp., but with **long flower stalks** to 70 mm, bearing **solitary pale blue to purple flowers** to 13 mm. Pod to 4 mm wide, scarcely veined. Sandy and rocky habitats.

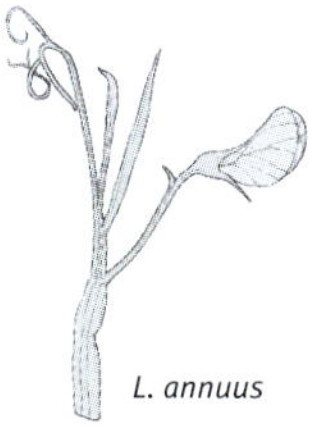
L. annuus

Lathyrus annuus ANNUAL YELLOW VETCHLING Similar to other *Lathyrus* spp. Leaves with 1 pair of lanceolate leaflets, arrow-shaped stipules and branched tendrils. Flowers borne in erect, long-stalked racemes of 1–3, **corolla yellow to orange** and red-veined. Calyx teeth equal, pod pale brown and glandular when young. Disturbed habitats.

P. sativum

Pisum sativum WILD PEA A variable, medium-tall, clambering, hairless annual to 2 m, stems **not winged**. Leaves with 1–3 pairs rounded to elliptical leaflets, more or less heart-shaped at the base, tendrils branched. Flowers white to purple, to 35 mm, borne in 1–3-flowered racemes, **wing petals fused to the keel**. Pod hairless, net-veined. Common in disturbed habitats and thickets.

O. reclinata

Ononis reclinata A low, slender, **spreading annual** with **shaggy-hairy stems** shorter than 15 cm. **Flowers small** (less than 10 mm), pink and solitary forming loose, leafy inflorescences; corolla equalling or shorter than the calyx; calyx with entire teeth. Local in coastal environments. **Ononis dentata** is very similar to *O. reclinata* but with 3-toothed or lobed calyx teeth. Local on maritime shale slopes and cliffs.

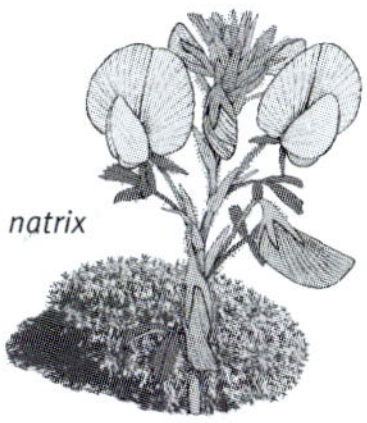
O. natrix

Ononis natrix LARGE YELLOW RESTHARROW A much-branched, cushion-like, lush green, **sticky subshrub** to 40 cm, stems woody below, densely leafy. Leaves trifoliate with ovate, toothed leaflets. **Flowers solitary** but borne in loose, leafy clusters, yellow, the standard petal veined with red externally, to 20 mm. Pod to 25 mm, pendulous and hairy. Very common to abundant throughout the region, especially near the coast. **Subsp. *ramosissima*** (sometimes treated as a true species) is the common coastal form in the region, distinguished by the **short glandular stem hairs**, and **calyx teeth just exceeding the length of their tube**.

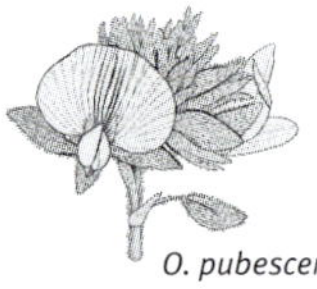
O. pubescens

O. biflora

Ononis viscosa Similar to *O. natrix* but a *taller*, softly hairy annual to 80 cm with a pinkish, glandular-hairy standard, and **3-veined calyx teeth.** ***Ononis pubescens*** is also similar to *O. natrix* but extremely sticky, only the middle leaves trifoliate, the others **1-foliate**. Flowers yellow, the standard red-veined, the calyx teeth **5-veined**. ***Ononis biflora*** is an erect annual distinguished by its somewhat fleshy, sparsely hairy leaves, and **1–2-flowered primary branches of the inflorescence** (not always 1-flowered), and pink-veined yellow, **hairless flowers**. ***Ononis maweana*** is very similar to *O. biflora* but with flowers with a white, pink or purple and glandular-hairy standard, and yellow wings and keel; on sandy soils. ***Ononis variegata*** is a **low, spreading, mat-forming annual**; densely hairy with bluish-green leaves and yellow flowers. Local on sand dunes.

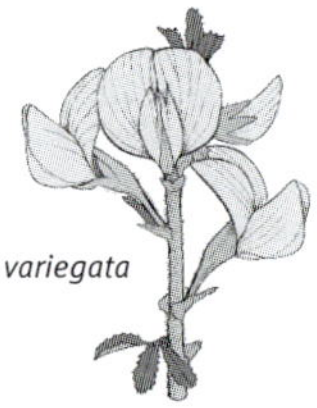
O. variegata

O. spinosa

Ononis spinosa **SPINY RESTHARROW** A dwarf-shrub, similar in stature to *O. natrix* but very spiny, to 80 cm, with 2 opposite rows of hairs on young stems. Leaves trifoliate, or simple above. Flowers large, pink, to 20 mm, usually borne singly at each node, forming a loose inflorescence. Pod hairy, usually with a single (up to 4), warty seed(s). Dry, rocky and waste places. ***Ononis diffusa*** is a similar but smaller, very sticky annual to 40 cm. Leaves trifoliate, with **toothed margins**. Flowers small to 11 mm, corolla exceeding the calyx; pink with a whitish keel, borne singly at each node in a **dense, oblong spike**. ***Ononis mitissima*** is a similar but rather taller and more erect, more or less hairless annual, to 60 cm, with trifoliate bracts with conspicuous white stipules. Flowers pink and exceeding their calyces. ***Ononis baetica*** is very similar to *O. mitissima* but with a **corolla that equals the calyx**; sandy and grassy places.

O. mitissima

Melilotus indica **SMALL MELILOT** An erect, branched or simple annual to 50 cm with trifoliate leaves, with toothed leaflets and more or less untoothed stipules. Flowers very small to 3 mm, pale yellow and borne in dense, **many-flowered racemes**; wings and keel equal, and shorter than the standard. Pod more or less spherical, pale brown when ripe. Very common in waste places, disturbed ground and on damp, sandy soils throughout. ***Melilotus italica*** is similar to *M. indica* but with larger flowers to 9 mm long, and pods that are wrinkled and with small depressions when ripe; dry open habitats.

M. elegans

Melilotus elegans An annual similar to *M. indica* with a stem hairy above, and small flowers to 3 mm with the **standard and wing petals shorter than the keel**, borne in 15–30-flowered racemes. Pod ovate, compressed and with very prominent transverse veins. ***Melilotus segetalis*** is similar to *M. elegans* but with **larger flowers to 8 mm**, the standard shorter than the keel, and a **yellow, egg-shaped pod** with concentric grooves. Damp habitats. ***Melilotus sulcata*** is very similar to *M. segetalis* but with small flowers to 4 mm, and a more or less **stalkless, almost globose pod**.

M. messanensis

Melilotus messanensis Similar to *M. segetalis* with yellow flowers to 5 mm, more or less equal standard and keel that exceed the wings, borne in 3–10-flowered **racemes that are much shorter than the subtending leaf**. Legume yellowish with concentric grooves. Cultivated and damp maritime places.

M. neapolitanus

Melilotus neapolitanus A short annual to less than 50 cm, hairy above with yellow flowers to 6 mm with equal wings and keel, borne in **short, lax racemes** of 8–20 flowers, to 10 mm long. Pod tiny, to 3 mm, globose, not flattened and with a short beak. In dry, sandy places.

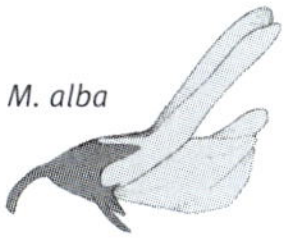

M. alba

Melilotus alba Similar to other *Melilotus* spp. but distinguished by its white flowers; common in a range of dry, disturbed habitats.

Trigonella foenum-graecum CLASSICAL FENUGREEK A short to medium, more or less hairy annual to 50 cm. Leaves trifoliate, oblong, toothed near the tip. Flowers cream, flushed with purple at the base, to 18 mm long, solitary or paired in the axils of upper leaves. Pod linear and held erect, to 10 cm long with a long beak, hairless, with longitudinal veins. Cultivated and sometimes naturalised.

M. lupulina

Medicago lupulina BLACK MEDIC A spreading, hairy annual with leaves with rounded to diamond-shaped trifoliate leaflets, often notched at the tips. Stipules toothed or entire. Flowers small, to 30 mm, yellow, and borne in ball-like, dense clusters of up to 50 flowers. Pod coiled, black when ripe and net-veined. Common in waste places.

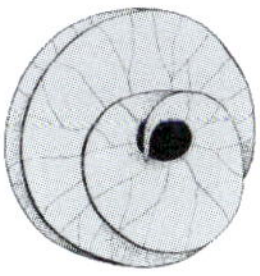

M. sativa

Medicago sativa LUCERNE A variable, hairy perennial to 80 cm. Leaflets oblong to linear, toothed at the apex. Stipules toothed at the base. Flowers violet or blue, borne in rounded, dense racemes of up to 50 flowers. Pod spiralled with 1–3 turns and a hole through the centre, not spiny. Common on waste places and on roadsides.

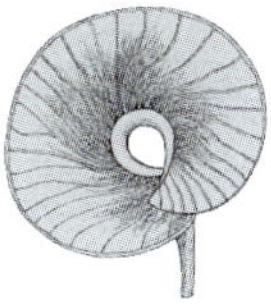

M. arborea

Medicago arborea TREE MEDIC A silvery-grey leafy **shrub** to 4 m with silky-white younger branches. Leaflets narrowed at the base, slightly toothed at the apex. Stipules untoothed. Flowers yellow, to 15 mm, borne in dense racemes of up to 8. Pod thin and coiled with 1 turn, hairless and net-veined. In rocky habitats or naturalised from gardens.

1. *Ononis pubescens*
2. *Melilotus indica*

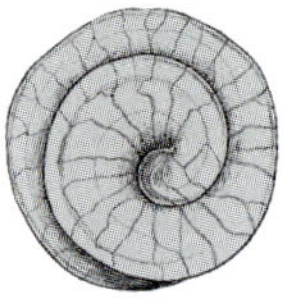

M. orbicularis

Medicago orbicularis **LARGE DISC MEDIC** A low to short, hairless or slightly hairy annual to 90 cm long. Leaflets oval-wedge-shaped, toothed at the apex. Stipules deeply toothed. Flowers to 5 mm, yellow, in racemes of up to 5. **Pod large and disc-like,** to 17 mm, smooth and spiralled anticlockwise in 4–6 turns, without a central hole. Frequent in waste places.

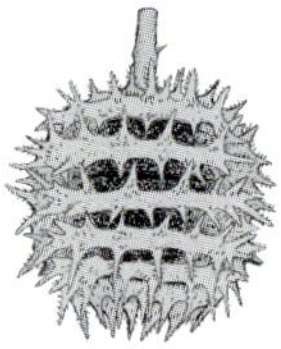

M. ciliaris

Medicago ciliaris A low to short annual with spreading stems. Leaflets sometimes with a dark spot and finely toothed. Stipules toothed. Flowers dull yellow, to 9 mm, borne in clusters of up to 7. **Pod glandular-hairy and spiny**, large, to 17 mm, with up to 10 anticlockwise spirals. Damp waste places. ***Medicago intertexta*** is similar but with **hairless pods** with longer **spines to 4 mm, appressed to the pod**. Habitats similar to those of *M. ciliaris*.

1

2

3

4

5

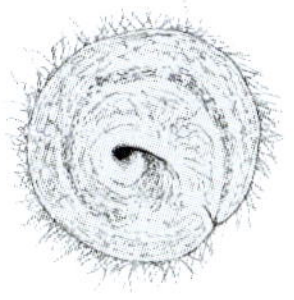

M. marina

Medicago marina SEA MEDIC A creeping, prostrate **white-downy perennial.** Leaflets ovate, toothed at the apex; stipules toothed or not. Flowers bright yellow, to 8 mm in short clusters of 5–12. Pod to 7 mm with 2–3 anticlockwise spirals and a hole in the centre, white-downy with 2 rows of spines. Common on coastal sands and dunes.

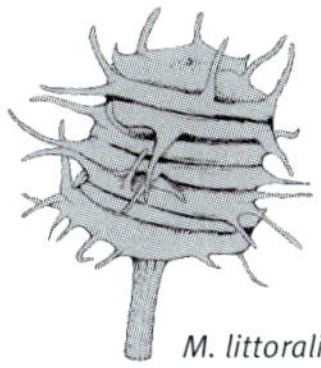

M. littoralis

Medicago littoralis A low to short, spreading, hairy annual with purplish stems. **Leaflets hairy on both surfaces**, oval to heart-shaped, toothed at the apex. Stipules toothed. Flowers yellow to orange, to 6 mm long in clusters of up to 6. Pod to 6 mm, disc-like to cylindrical with 3–6 turns, spiny and hairless, with a groove along the margin. Coastal habitats. **Medicago aculeata** is similar to *M. littoralis* but densely hairy and with a larger pod (to 10 mm) clothed in **short, dense hairs**, with or without spines. **Medicago blancheana** is like *M. littoralis* but with deeply toothed leaflets, and pods not spiny or hairy. Cultivated land. **Medicago truncatula** is similar to *M. littoralis*, with often black-spotted leaves, a more robust stature, and with clusters of up to 3 flowers only. Pods **with a 3-keeled margin** and slender spines. Rocky inland slopes.

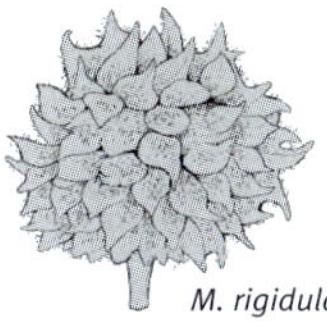

M. rigidula

Medicago tornata Similar to *M. truncatula* but with more than 3 flowers in a cluster, and a pod with up to 8 spirals, usually without spines, or with short, straight spines.

M. turbinata

Medicago rigidula A glandular-hairy annual rather similar to *M. truncatula*, characterised by its spiny and **densely glandular-hairy pods** with 4–7 anticlockwise turns. Common on roadsides, fallow land and in grassy places.

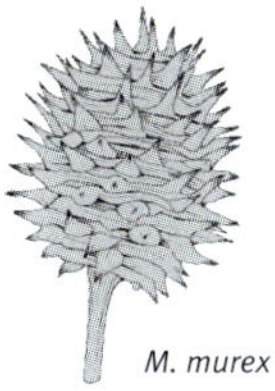

M. murex

Medicago turbinata A robust, prostrate or ascending, hairy annual with reddish stems and toothed, trifoliate leaves. Flowers yellow-orange, to 6 mm long, borne in clusters of 3–8. Pod barrel-shaped, to 7 mm, with up to 6 clockwise or anticlockwise turns, a single groove along the margin, and **short, broad spines**. In various disturbed habitats. **Medicago murex** is rather similar but with pods with 5–9 anticlockwise turns, with 3 grooves along the margin, and **spines long and pointed**. Habitats similar to those of *M. turbinata*.

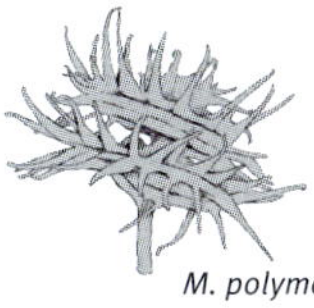

M. polymorpha

M. minima

Medicago polymorpha A very variable, low, prostrate and spreading hairy or hairless annual with purplish stems and trifoliate leaves, toothed at the apex. **Stipules deeply toothed with slender teeth**. Flowers yellow, to 4 mm long, borne in clusters of 1–5. Pod disc-like or cylindrical to 8 mm with 1–6 (usually 4) anticlockwise turns, usually spiny with short, hooked spines (rarely only warted), the outer margin with 2 rows of spines; **pod globose but distinctly flattened** above and below, and netted (**not shiny**) on the surface. Common on roadsides, grassy places and cultivated fallow land. **Medicago minima** is similar and rather easily confused with *M. polymorpha* but often is a more densely hairy plant with ball-like, spiny pods to 5 mm (rather large relative to the plant), with 3–5 turns and long, slender spines that are weakly hooked; not distinctly flattened, and often rather shiny (less netted) on the surface. Habitats similar to those of *M. polymorpha*.

---

1. *Medicago orbicularis*
2. *Medicago marina*
3. *Medicago littoralis*
4. *Medicago polymorpha*
5. *Medicago minima*

M. arabica

Medicago arabica SPOTTED MEDIC A low, prostrate and spreading more or less hairless annual with trifoliate leaves. Leaflets notched and with a **conspicuous dark spot**. Flowers yellow, to 7 mm borne in racemes of up to 4. Pod cylindrical, to 6 mm, hairless, and with up to 7 anticlockwise turns with **spreading spines** and a margin with 3 grooves. Common on fallow land and in pine forests.

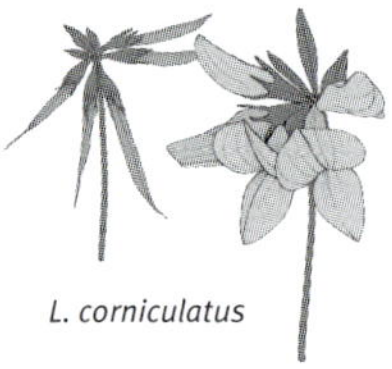

L. corniculatus

Lotus corniculatus COMMON BIRD'S-FOOT TREFOIL A sprawling or creeping, almost hairless perennial with a **woody stock**. Pinnate leaves that appear trifoliate because the lowermost pair is at the base of the stalk; leaflets ovate, lanceolate or round. Flower-heads with 2–8 yellow-orange flowers (red in bud) on long stalks to 80 mm long. Pods 15–30 mm long. Grassy places.

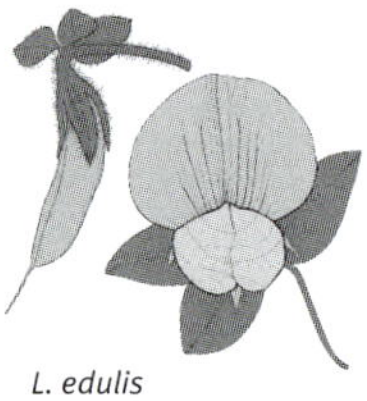

L. edulis

Lotus creticus SOUTHERN BIRD'S FOOT TREFOIL A low, spreading, silvery-hairy perennial with leaves with oblong leaflets. Flowers yellow with a purple-tipped keel, borne in long-stalked racemes; calyx distinctly 2-lipped. Pod straight or curved, many-seeded and to 50 mm. Common in sandy maritime habitats.

Lotus edulis EDIBLE LOTUS A low to short, spreading, slightly hairy annual with ovate leaflets. Flowers yellow with a purple-tipped keel, to 18 mm long, borne in heads of 2–6; calyx bell-shaped. Pod oblong and curved, **very inflated**, with a groove along the back, to 40 mm long. *Matos* and roadsides.

L. conimbricensis

Lotus conimbricensis A more or less hairless annual with ovate to diamond-shaped leaflets. Flower stalks shorter than the leaves, with solitary flowers; **corolla white to pink** with a violet keel. Legume slender, to 60 mm and curved upwards. Dry grassy places and dunes.

L. ornithopodioides

Lotus ornithopodioides A low to short, spreading, hairy annual with ovate to diamond-shaped leaflets, the lower two triangular and heart-shaped at the base. Flowers yellow, to 10 mm long, borne in heads of 2–5 on stalks slightly longer than the leaves (at least in fruit); calyx 2-lipped. Pod to 50 mm long, **constricted between the seeds**. Damp habitats near water.

Ornithopus pinnatus ORANGE BIRD'S FOOT A low, spreading, prostrate, hairy annual. **Leaves pinnately lobed** with 3–7 pairs of leaflets. **Flowers orange-yellow**, borne in heads of 3–5 without a bract below. Pod linear, to 50 mm long, flattened and markedly constricted between the seeds. Grassy and bare, stony habitats. ***Ornithopus perpusillus*** has pinnately lobed leaves with 4–14 leaflets and a terminal leaflet. Flower-heads with a pinnate bract below; **corolla cream to white** with red veins. **Pods curved and spreading**, in clusters of usually 2–3, and constricted between the seeds.

Ornithopus compressus is a low, spreading, hairy annual with pinnate **leaves with up to 18 pairs of leaflets**. Flowers yellow, borne in heads of 3–7, **subtended by a leafy bract**. Pod linear, flattened and curved, slightly constricted between the seeds. Common in grassy places. ***Ornithopus sativus*** is similar but has **pink-white flowers** and pods constricted between the seeds. Occasionally naturalised in grassy waste places.

1. *Medicago arabica*
2. *Lotus corniculatus*
3. *Lotus creticus*
4. *Lotus conimbricensis*
5. *Ornithopus compressus*

Trifolium campestre HOP TREFOIL A hairy annual to 30 cm. Leaves alternate, the terminal leaflet longer-stalked than the laterals. **Flowers yellow**, to 5 mm long, borne in **rounded heads to 15 mm across**. Common in a range of habitats in the region.

Trifolium pratense RED CLOVER A perennial, initially rosette-forming, later spreading; rather hairy. Leaflets circular to ovate, more hairy below than above. **Flower-heads large, to 30 mm across**, spherical or ovoid, solitary or paired, and usually **stalkless**; calyx tube 10-veined with triangular teeth that are bristle-pointed at the tips; corolla to 15 mm, usually pink, rarely cream or white.

T. suffocatum

Trifolium suffocatum SUFFOCATED CLOVER A distinctive **prostrate**, tufted, low annual to less than 50 mm high. Leaves with oval leaflets overtopping the **unstalked** rounded flower-heads. Flower-heads terminal and densely crowded; corolla white; calyx tube nearly cylindrical. Local on coastal sands and bare ground.

1. *Trifolium campestre*
2. *Trifolium pratense*
3. *Trifolium repens*
4. *Trifolium resupinatum*
5. *Trifolium resupinatum*
6. *Trifolium angustifolium*
7. *Trifolium stellatum*
8. *Trifolium stellatum*

T. isthmocarpum

Trifolium repens WHITE CLOVER A variable low to short, creeping perennial with stems rooting at the nodes, often hairless. Leaves trifoliate with green, ovate to elliptical leaflets that are paler along the veins. Flowers normally white, sometimes pink or reddish, to 10 mm long, borne in dense, long-stalked globose heads; calyx tube longer than wide; flowers scented. Pod linear and compressed between the seeds. Cultivated for fodder, and common in grassy places.

Trifolium isthmocarpum A low to medium, spreading or ascending hairless annual. Leaves trifoliate with oval to elliptical leaflets; stipules membranous and pointed. Flowers pink or more rarely white, to 12 mm long, borne in **large cylindrical or hemispherical heads** to 25 mm across, with stalks exceeding the leaves. Pod oblong and compressed between the seeds.

T. resupinatum

Trifolium resupinatum REVERSED CLOVER A low to medium, spreading or sprawling, more or less hairless annual. Leaves trifoliate with wedge-shaped leaflets. **Flowers borne upside down** in circular, flattened heads, pink to purple, 5–8 mm long, on rather short stalks; calyx conspicuously inflated and papery in fruit, rather hairy. Common in moist, sandy or grassy habitats. **Trifolium tomentosum** is similar to *T. resupinatum*, but with smaller flowers to 4 mm long and spherical fruiting heads **clothed in soft, white hairs** appearing as cotton-like balls, almost stalkless.

T. spumosum

Trifolium spumosum A spreading to erect annual to 30 cm, rather similar to *T. resupinatum*. Leaves trifoliate with wedge-shaped, toothed leaflets that are not strongly veined. Flower-heads more or less spherical, borne on short stalks; corolla pink, slightly exceeding the calyx; **calyx hairless** and inflated in fruit, forming somewhat spiky fruiting heads.

Trifolium angustifolium NARROW-LEAVED CRIMSON CLOVER A short to medium, somewhat hairy annual. **Leaves with linear-lanceolate**, pointed leaflets. Flowers pink, to 12 mm, borne in stalked, **cylindrical heads**, opening from the top downwards; corolla not (or scarcely) exceeding the calyx. Common in a range of disturbed or sandy habitats.

T. stellatum

Trifolium stellatum STAR CLOVER A low to short, erect, hairy annual with stems simple or branched from the base. Leaflets oval and slightly toothed with ovate, **toothed stipules** with bright green veins. Flowers pink or yellowish, to 12 mm borne in solitary, large spherical heads to 25 mm across in fruit; calyx equalling the corolla, densely white-hairy with slender, reddish, **spreading lobes** (star-like). Very common in a range of disturbed and grassy habitats.

T. hirtum

Trifolium hirtum HAIRY TREFOIL A short, spreading, **hairy**, branched annual. Leaflets wedge-shaped and finely toothed towards the tip. **Stipules long and straight** with a hairy tip. Flowers pink to purple, to 15 mm long, borne in large, **densely hairy heads** to 20 mm wide with a pair of leaves immediately below; **corolla greatly exceeding the calyx**. Dry, stony habitats. **Trifolium scabrum** is similar to *T. hirtum* but with smaller, **creamy white flowers** to 5 mm long that turn pink with age and equal or slightly exceed the calyx, borne in unstalked heads to 10 mm across. Habitats similar to those of *T. hirtum*. ***Trifolium cherleri*** is similar to *T. hirtum* but rarely branched above, with heart-shaped more or less untoothed leaflets, and the **corolla equalling or shorter than the calyx**. ***Trifolium lappaceum*** is similar to *T. hirtum* but with a hairless calyx, and short-stalked flower-heads.

Trifolium leucanthum A densely hairy annual distinguished by its wedge-shaped leaflets and white or pink flowers to 8 mm, in which the corolla equals (not exceeds) the calyx; **calyx with equal, lanceolate, 3-veined teeth**. Flowers borne in more or less spherical heads, often in pairs, on stalks to 12 cm. Dry places and pine woods.

T. squamosum

Trifolium squamosum An erect or ascending more or less hairy perennial with elliptical leaflets; **stipules long and spreading**. Terminal flower-heads becoming ovoid, short-stalked and pink-white and rather small relative to leaves; each flower-head subtended by a pair of leaves; calyx teeth unequal, the corolla longer than the calyx. **Fruiting heads resemble miniature teasels** by virtue of spreading calyx teeth. Rare in maritime habitats. Grows in grassy, saline waste areas around Lagos.

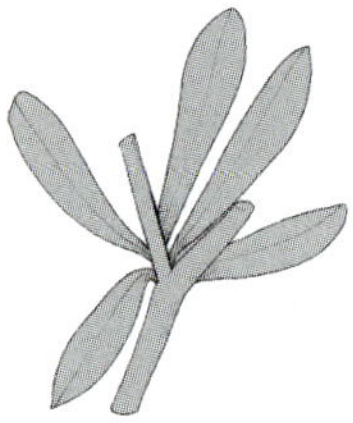

D. pentaphyllum

Dorycnium hirsutum DORYCNIUM A sprawling perennial to 50 cm, **densely woolly hairy**. Leaves pinnate with 5 leaflets, the lowermost pair stipule-like, the true stipules minute. Flowers with a **corolla 10–20 mm**, pale pink or white, with a dark red or blackish keel, borne in a compact, stalked raceme of 4–10 flowers. Pod oblong, to 12 mm. Common on the *matos* and on sand dunes. ***Dorycnium pentaphyllum*** is very similar to *D. hirsutum* but taller, to 80 cm, leaflets all arising from a single point (without a rachis). **Flowers small, 3–6 mm**, white with a red-black keel. Pod egg-shaped, to 5 mm. Habitats similar to those of *D. hirsutum*. ***Dorycnium rectum*** is similar to *D. hirsutum* with pinnate leaves (the leaflets not arising from a single point), but less hairy, with **smaller flowers, the corolla 3–6 mm**, white with a purplish keel. Damp and aquatic habitats.

Anthyllis tetraphylla (= Tripodion tetraphyllum) BLADDER VETCH A low, spreading, hairy annual. Leaves normally with 5 leaflets, the terminal lobe largest and ovate. Flowers yellow with darker wings and a red-tipped keel, to 12 mm long, borne in dense lateral clusters of 1–7; calyx with silvery hairs, becoming very inflated and bladder-like in fruit, pale green tipped with red. Very common in a range of dry and disturbed habitats.

Anthyllis vulneraria MEDITERRANEAN KIDNEY VETCH A low, tufted, hairy perennial. Lower leaves with a single leaflet, upper leaves with 7–13 elliptical leaflets. Flowers red or purple, to 15 mm long borne in long-stalked heads with a pair of **leaf-like bracts** beneath; calyx shiny with silky hairs, purple-tipped. Very common in a range of dry habitats, often coastal.

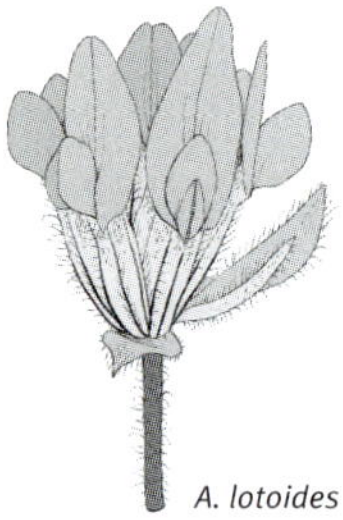

A. lotoides

Anthyllis lotoides (= Hymenocarpos lotoides) An erect, hairy annual. Lower leaves with a single terminal leaflet, the upper leaves with 3–7 more or less equal, narrowly lanceolate leaflets that are hairy on both sides, and especially on the margins. Flower-heads with up to 8 flowers with bracts immediately beneath; calyx tubular with 5 unequal teeth and covered in long hairs; **corolla bright yellow-orange**. **Legume straight,** with up to 10 seeds. Local and sporadic on fixed dunes and cliff tops in the region.

A. cornicina

Anthyllis cornicina An erect hairy annual rather similar to *A. lotoides*. Lower leaves with a single terminal leaflet, the upper leaves with up to 9 more or less equal, **narrow lanceolate leaflets** that are silky-hairy on both sides. Flowers borne in heads of up to 9; calyx not inflated in flower, but inflated in fruit, with 5 unequal teeth; corolla bright yellow-orange. **Legume winged, curved** into an almost complete circle.

1. *Trifolium squamosum*
2. *Dorycnium hirsutum*
3. *Anthyllis tetraphylla*
4. *Anthyllis vulneraria*
5. *Anthyllis lotoides*

1. *Coronilla valentina*
2. *Scorpiurus muricatus*
3. *Scorpiurus muricatus*
4. *Hedysarum glomeratum*

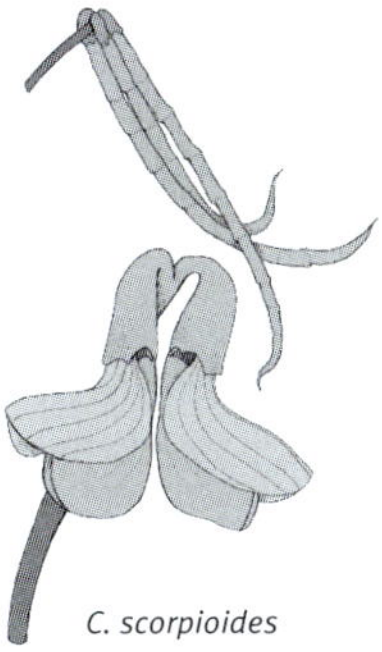

C. scorpioides

Coronilla valentina (= Coronilla glauca) A blue-green dwarf shrub to 1 m, with pinnate leaves with 3–6 pairs of notched, elliptical, **not fleshy leaflets**. Flowers borne in lateral clusters of 4–12, yellow, strongly scented; calyx bell-shaped. Pod to 50 mm long, with 3–7 swollen segments. Common on scrubby cliff tops. ***Coronilla juncea*** is similar to *C. valentina* but with green, **rush-like stems** with long internodes and few leaves. Leaves with 2–3 pairs of round to elliptical **fleshy leaflets** that are soon-falling. Less common than *C. valentina* but in similar habitats.

Coronilla scorpioides **ANNUAL SCORPION VETCH** A prostate to almost erect, short, hairless, blue-green annual. Lower leaves simple or trifoliate, the **terminal leaflets elliptical to rounded, much larger than the other leaflets**; upper leaves simple or trifoliate. Flowers yellow, often with brownish veins, to 8 mm long borne in stalked heads of 2–5. Pod ridged, to 60 mm, curved, with up to 11 jointed segments. Locally common in dry, open habitats. ***Coronilla repanda*** is similar to *C. scorpioides* but with **upper leaves pinnate** with a single terminal leaflet, flowers that become reddish on drying and a **curved, segmented pod**. Coastal habitats.

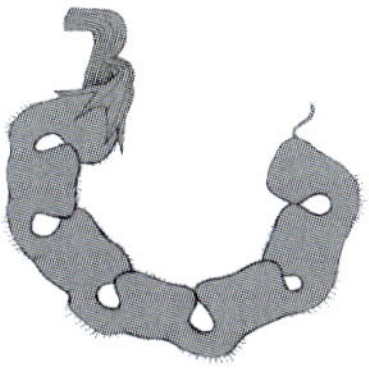

H. ciliata

Coronilla minima* subsp. *minima is a low to short shrub with pinnate leaves with 2–6 pairs of ovate **unstalked leaflets** and small stipules to only 1 mm long. Corolla 5–8 mm long, yellow. Local in exposed habitats.

Hippocrepis ciliata A slender, annual herb to 30 cm with pinnate leaves with 3–7 pairs of linear to ovate leaflets. Flowers borne in clusters of 2–6 on **stalks that equal the leaves**; corolla yellow, to 5 mm long. Pod flattened and curved upwards almost into a full circle, with very pronounced segments, and hairs on 1 side. Local on coastal sands, dry slopes and pine forests. ***Hippocrepis multisiliquosa*** is similar but with larger flowers to 8 mm long that are normally solitary, and the pod **curved downwards** into an almost complete circle; **hairless**. Habitats similar to those of *H. ciliata*.

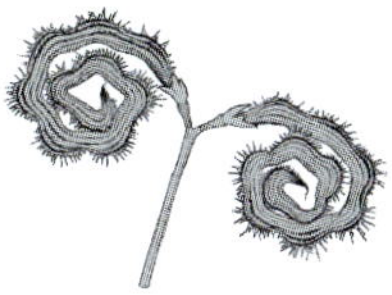

S. muricatus

Scorpiurus muricatus A low hairy or hairless sprawling annual with simple, elliptical, entire leaves with 3 prominent veins; the upper leaves short-stalked or unstalked. Flowers yellow, sometimes flushed with red, to 10 mm long, borne in long-stalked clusters of 2–5. Pod to 50 mm long, coiled and twisted (so appearing shorter), covered in short, robust hairs. Common in dry, rocky and sandy habitats. ***Scorpiurus vermiculatus*** is similar but with **solitary flowers** (rarely 2 together) to 20 mm long, and the **fruit with warts not hairs**. Habitats similar to those of *S. muricatus*.

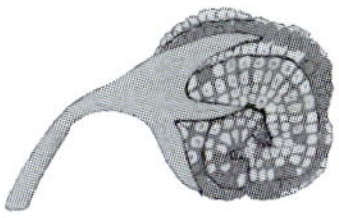

S. vermiculatus

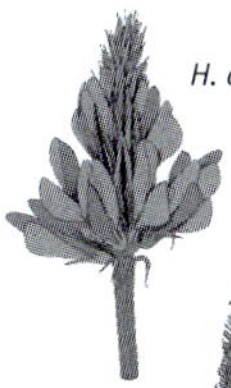

H. coronarium

H. glomeratum

Hedysarum coronarium **ITALIAN SAINFOIN** A robust, rather hairy perennial. Leaves pinnate with up to 5 pairs of elliptical to rounded leaflets that are hairy beneath, and untoothed. **Flowers magenta**, to 15 mm long, borne in conspicuous, large, dense and long-stalked racemes of up to 45 flowers. Pod with up to 4 spiny segments; hairless. Cultivated for fodder and naturalised on disturbed and fallow ground. ***Hedysarum flexuosum*** is similar but with leaves with **1–3 pairs of leaflets**, the terminal leaflet largest, and **smaller pink to purple flowers** to 12 mm long. Pod with up to 8 segments. Maritime habitats. ***Hedysarum glomeratum*** has ovate leaflets, bright pinkish-purple flowers borne in **small racemes of 2–10 flowers**, the corolla greatly exceeding the calyx. **Pod with hooked, purplish spines**. Dry, sandy places and pine forests. Common on sea cliffs in Sagres.

O. humilis

***Onobrychis humilis* (= *Onobrychis peduncularis*)** A hairy perennial to 60 cm. Pinnate leaves with up to 14 pairs of elliptical leaflets. Flowers borne in racemes; corolla bright pale pink with **prominent purple, longitudinal veins**. Pod not jointed, pale yellow-green with distinct reddish spines, and white woolly hairy. Local in sandy places.

Wisteria floribunda A vigorous, deciduous climber to 15 m tall with twining stems. Leaves alternate and pinnate with up to 19 leaflets. Flowers lilac and pea-like, borne in long, tapered racemes to 1 m long. Pod oblong and velvety. Native to Japan but widely planted.

POLYGALACEAE

Herbs and small shrubs with simple, usually alternate, leaves. Flowers cosexual, irregular, in slender racemes or spikes. Sepals 5, corolla with 3 fused petals, stamens 8. Fruit a 2-lobed capsule.

P. monspeliaca

Polygala monspeliaca An erect, low to short annual with large linear-lanceolate to lanceolate leaves to 25 mm long. Flowers whitish with green-white wings up to twice as long as the petals, keel with a large crest. Inflorescence somewhat asphodel-like at a glance. Local on dry and wooded slopes.

Order: ROSALES

ROSACEAE

A large family of herbs, shrubs and trees. Leaves alternate, simple or compound, and with stipules. Flowers often 5-parted, and with numerous stamens and few to many carpels. Fruit variable ranging from a capsule to an edible fruit.

Rosa pouzinii A lax shrub to 3 m with spines on the stems. Leaflets hairless or finely hairy on the underside; petiole with some glands. Flowers pale pink or occasionally white with 5 petals and numerous yellow stamens; **flowering stem with many glands**. Fruit characteristically rosehip-like: red, and ovoid, to 20 mm long.

S. minor

Sanguisorba minor **SALAD BURNET** A variable, greyish, bushy perennial with densely hairy stems. Leaves forming a basal rosette, pinnate with up to 12 pairs of elliptical, toothed leaflets. Flowers tiny, borne in egg-shaped heads with the upper flowers female with reddish styles, the lower male with yellow anthers; sepals bright green; petals absent. Fruit 4-angled and ridged. Very common in grassy places and fallow land throughout the region.

Rubus ulmifolius A bramble-like, spreading perennial with robust, thorny, arching or prostrate stems that are angled and grooved; hairy or not. Leaves often small with 3–5 leaflets, **leathery dark green above and shortly white-hairy beneath**, variously toothed and ovate in shape. Inflorescence long and narrow, and leafy at the base, with hairy prickles; sepals deflexed after flowering, and white-hairy; petals crumpled and pink or white with **stamens equalling or scarcely exceeding the styles**. Fruit a blackberry. Local in thickets, often on higher ground inland.

1. *Onobrychis humilis*
2. *Onobrychis humilis* fruits
3. *Polygala monspeliaca*
4. *Sanguisorba minor*
5. *Sanguisorba minor*

C. monogyna

Crataegus monogyna HAWTHORN A thorny shrub or small tree to 10 m forming a dense crown above. Spines to 20 mm, leaves oval to wedge-shaped, with 3–7 lobes darker above, toothed or not. Flowers white, 8–15 mm and borne in clusters; 1 style. Fruit a bright red berry. Common in thickets and hedges and on roadsides inland.

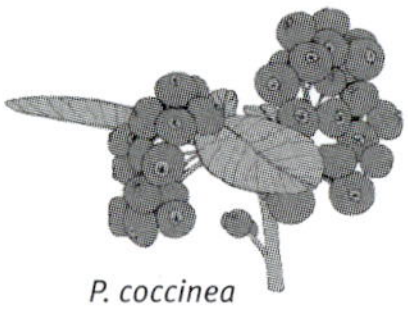

P. coccinea

Pyracantha coccinea FIRE THORN A small, very densely branched, spiny evergreen shrub to 2 m. Leaves elliptical, deep green and blunt-toothed, more or less hairless or hairy beneath when young. Flowers cream-white, 8 mm across. Berry orange or red with small persistent sepals. Cultivated and common as a hedge plant.

1. *Prunus dulcis*
2. *Prunus persica*
3. *Eriobotrya japonica*
4. *Eriobotrya japonica*

P. persica

Prunus dulcis **ALMOND** A deciduous shrub or small tree to 8 m, densely branched and spiny. Leaves oblong-lanceolate, hairless and toothed. Flowers appear before the leaves, pink in bud, fading to white, to 50 mm, borne in pairs. Fruit oval and laterally compressed, grey-green to yellow and velvety, to 60 mm. Very common in cultivation and on the *matos*. ***Prunus persica*** is a similar small tree to 6 m with **velvety** (hairless in var. *nucipersica*, the nectarine) yellow, orange or red**, large rounded fruits** (peaches) with solitary **flowers that remain mid-dark pink**; cultivated for peaches. ***Prunus armeniaca*** has broader, almost round leaves on stalks to 40 mm long, **flowers always pale pink-white**; cultivated for apricots.

P. lusitanica

Prunus lusitanica **PORTUGAL LAUREL** A dense, evergreen shrub to 15 m with leathery, **shiny green elliptical-ovate leaves** and numerous whitish flowers borne in **long erect racemes** exceeding the leaves. Fruit ovate, purplish black when ripe. Planted; local. ***Prunus laurocerasus*** is similar but with pale green twigs and petioles, leaves lighter green, and racemes scarcely exceeding the adjacent leaf. Planted and perhaps naturalised.

The following ***Prunus*** species are widely cultivated and possibly naturalised in the region:

P. cerasifera

Prunus cerasifera **CHERRY PLUM** A shrub or tree to 8 m with leaves markedly wavy-margined or toothed with forward-pointing lobes. Flowers white, mostly solitary, borne with or earlier than the leaves. Fruit a round, red or yellow berry to 30 mm across. Planted.

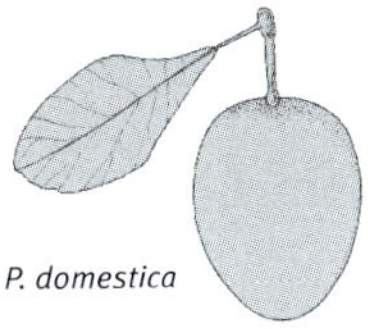
P. domestica

Prunus domestica **PLUM** A shrub or tree to 10 m with dull brown bark, and spiny branches in naturalised populations. Leaves dull green and hairless above, variably hairy beneath. Flowers white. **Fruit a hanging** round purple, red, green or yellow berry to 60 mm across (a plum). Planted.

Cydonia oblonga **QUINCE** A shrub or tree to 6 m with shoots hairy when young, later hairless. Leaves ovate and entire. Flowers to 45 mm across, borne on short, hairy stalks; petals pink, exceeding the sepals. Fruit many-seeded, yellow, pear-like and fragrant (a quince). Cultivated and in thickets.

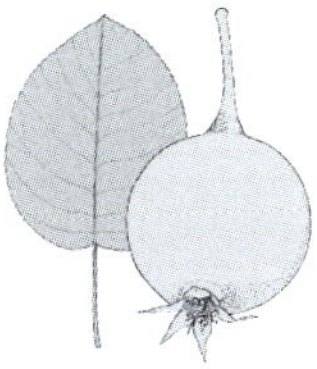
P. bourgaeana

Pyrus bourgaeana **WILD PEAR** A tree to 10 m tall with an open crown and spiny lower branches. Branches grey and stout. Leaves ovate to heart-shaped, thick, wavy-margined, hairless, at least when mature, with long, slender stalks to 50 mm long. Flowers white with rather long petals to 10 mm, exceeding the sepals. Fruit to 25 mm in diameter, fig-shaped, and dull greenish-yellow, later brown. Local, beside seasonal streams and rivers. ***Pyrus communis*** is the cultivated pear, and is distinguished from *P. bourgaeana* by its leaves that equal or exceed their **short stalks**, and its much larger, sweet-tasting pink-green fruit (a pear). Cultivated and naturalised in the region.

P. communis

E. japonica

Eriobotrya japonica **JAPANESE LOQUAT** A very robust small tree to 10 m. Young stems covered in red-brown felted hairs. Leaves elliptical, large, to 30 cm long, dark green and shiny above, felted below. Flowers white, 10 mm borne in terminal panicles; 20 stamens. Fruit pear-shaped, to 60 mm, yellow when ripe with 1–3 large seeds. Widely cultivated as an ornamental and for its edible fruits in gardens throughout the area.

RHAMNACEAE

Trees or shrubs with simple leaves and with stipules. Flowers 4–5-parted, the petals small or absent, often hooded over the stamens. Fruit a fleshy berry.

R. alaternus

Rhamnus alaternus MEDITERRANEAN BUCKTHORN A dioecious, erect, superficially holly-like evergreen shrub to 5 m, not spiny and more or less hairless. **Leaves leathery**, alternate, ovate and shiny dark green, toothed or sparsely toothed. Flowers borne in dense, cylindrical racemes yellowish, borne in small clusters in the leaf axils, lacking petals, 5-parted. Fruit a berry, red ripening black. Very common on the *matos* throughout the region. *Osyris quadripartita* has a similar habit but has 3–4 parted flowers. ***Rhamnus lycioides*** is similar to *R. alaternus* but with **spines present** and **4-parted (not 5-parted) flowers**. Habitats similar to those of *R. alaternus*.

MORACEAE

A large, mainly tropical family including the fig, mulberry and breadfruit.

Ficus carica FIG A deciduous tree to 10 m with greyish branches and large, palmately lobed leaves. Flowers minute, borne within a syncarp. Fruit a fig; pear-shaped and purplish when mature and edible. Widely planted and naturalised on rocky slopes and near streams in the region; an important constituent of the Portuguese *matos*.

Ficus elastica RUBBER PLANT A tree with large, thick, leathery green or yellow-green leaves to 30 cm long. Commonly planted as an exotic ornamental in gardens, parks and on roadsides in the Algarve.

URTICACEAE

Herbs with opposite or alternate, simple leaves. Flowers normally unisexual and on the same or separate individuals, greenish and 4–5-parted with no petals.

Urtica dioica STINGING NETTLE A familiar, robust and usually dioecious perennial to 1 m tall, with stinging bristles. Leaves ovate, **cordate** at the base and toothed. Racemes to 10 cm, exceeding the petioles. Common in damp, disturbed habitats, grassy places and woods.

***Urtica dubia* (= *Urtica membranacea*)** MEMBRANOUS NETTLE A short to tall monoecious **annual** similar to *U. dioica*. Leaves with petiole about as long as the lamina; **stipules only 2 at each node**. Racemes long and conspicuous, to 10 cm long; lower racemes female, shorter than petioles, upper racemes male and longer than petioles, often coiling. Common in disturbed and grassy places with nutrient-rich soils.

Parietaria judaica PELLITORY OF THE WALL A short, tufted and spreading perennial with much-branched, reddish stems. Leaves oval, pointed, 10–70 mm. Flowers unisexual but borne on the same plant. A familiar plant on walls and damp rocks. Common in towns.

CANNABACEAE

A small family of plants including *Cannabis* (hemp) and *Humulus* (hops). Flowers unisexual, male and female on the same or separate plants; 5-parted.

Celtis australis **European nettle tree** A deciduous tree to 20 m tall with grey bark, simple, alternate, rough leaves and small, green cosexual flowers that lack petals. Fruit a small, purple, berry-like drupe borne in clusters.

1. *Rhamnus alaternus*
2. *Rhamnus lycioides*
3. *Ficus carica*
4. *Urtica dioica*
5. *Urtica dubia*
6. *Urtica dubia*

Order: **CUCURBITALES**

CUCURBITACEAE

Herbs with often climbing or trailing stems and tendrils, and alternative leaves. Flowers unisexual, usually 5-parted with a deeply lobed corolla; stamens usually 3. Fruit succulent and berry- or pod-like.

Ecballium elaterium **SQUIRTING CUCUMBER** A spreading, very bristly perennial without tendrils, with a tuberous rootstock. Leaves rough with bristles, more or less triangular and long-stalked. Flowers small (to 20 mm across) and pale yellow; male and female flowers borne separately. Fruit a bristly pod to 50 mm long borne on long stalks, **exploding violently** from the point of attachment when ripe. Common throughout the Mediterranean but local in the Algarve; on sandy waste ground.

Bryonia cretica* subsp. *dioica **WHITE BRYONY** A dioecious, climbing, hairy perennial with **coiled tendrils**. Leaves palmately 5-lobed and plain green. Flowers greenish-white with darker veins, to 18 mm across. Fruit a berry, green with white markings then red when ripe. Rather uncommon in the region; in woods and thickets.

1. *Ecballium elaterium*
2. *Castanea sativa*
3. *Quercus coccifera*
4. *Quercus ilex*

1

2

3

4

The following species are widely cultivated and possibly naturalised in the region for their ornamental or edible fruits:

Luffa aegyptiaca A climber with bright yellow flowers and large, hanging, cucumber-like fruits to 30 cm long.

Cucurbita moschata A vigorous trailing species with large, bristly, palmate **leaves with pale markings**. Flowers bright yellow. Fruit large, spherical, pumpkin-like, initially green with paler markings, ripening orange.

Citrullus lanatus **WATERMELON** A vigorous trailing species with large, bristly, palmate, **deeply divided leaves**. Fruits very large, dark green with paler markings.

Order: FAGALES

FAGACEAE

Trees or shrubs with alternate, simple leaves. Male flowers borne in catkins, female flowers separate but borne on the same individual. Fruit a nut.

C. sativa

Castanea sativa **SWEET CHESTNUT** A large, deciduous tree to 30 m, trunk with grey-brown bark, often with spiralled fissures. Leaves oblong-lanceolate, with pronounced veins and sharply toothed. Male flowers yellowish in long catkins (to 18 cm); female flowers few, on lower branches. Fruit a chestnut in a spiny, dehiscent husk. Local in woods on acid soils.

Quercus Evergreen or deciduous trees or shrubs with fruit a distinctive nut (acorn) with a cup (cupule).

Q. coccifera

Quercus coccifera **KERMES OAK** A dense, evergreen shrub to 2 m with scaly, grey bark. Leaves ovate-oblong with spiny teeth, short-petioled, leathery, and hairless when mature, with veins prominent above but not beneath; midrib straight. Fruit ripening in the second year with conspicuous scales on the cupule; solitary or in pairs. A red dye is derived from a scale insect that commonly infects this species. Common on the *matos* throughout the Algarve.

Q. ilex

Quercus ilex **HOLM OAK** A large evergreen tree to 25 m with downy young branches and grey bark. Leaves leathery, oblong to lanceolate, untoothed or sometimes spiny-toothed, **downy and with prominent veins beneath** when mature. Acorn bitter-tasting with dense scales on the cupule. Common in scrub and on the *matos*, often planted. ***Quercus rotundifolia*** is often treated as a subspecies of *Q. ilex* and differs in having broader leaves with fewer veins that are grey-glaucous above, and sweet-tasting acorns. Habitats similar to those of *Q. ilex*.

Q. faginea

Quercus faginea A semi-evergreen shrub or tree to 20 m. Leaves leathery, oblong or elliptical with large, broad and forward-pointing teeth; shiny above and **densely downy beneath**, acorns virtually stalkless. ***Quercus valentina*** is probably conspecific, but differs in being more shrubby, with smaller leaves with acute teeth and longer hairs on the lower surface.

Q. suber

Quercus suber **CORK OAK** A large evergreen tree to 20 m with **thick, corky and deeply ridged bark**, red beneath, and downy young branches. Leaves oblong, dark green above and grey and downy beneath, toothed; midrib sinuous (not straight). Fruit ripening in the first year, the cupule with long and spreading scales. Common on rocky hillsides; frequently cultivated for cork.

Q. canariensis

Quercus canariensis A very large semi-evergreen tree to 30 m with densely downy young branches, virtually hairless later. Leaves ovate-elliptical, large, to 15 cm long, with more or less acute teeth, **glaucous beneath**. Scales on cupule small and loosely appressed. Very rare, in cork oak forests; can be seen on roadsides near Monchique.

Quercus lusitanica A semi-evergreen tree or shrub to 20 m (often less). Leaves to 10 cm long, rather leathery, elliptical and sinuately toothed, hairless above and felted below and short-stalked. Acorns more or less stalkless, borne in downy cups with appressed scales. Local on hill slopes; common on Picota.

BETULACEAE

A family of deciduous trees or shrubs with alternate leaves. Male flowers hang in catkins, female flowers are held erect.

Alnus glutinosa **ALDER** A tree to 20 m with dark brown, fissured bark and hairless twigs. Buds purplish and short-stalked. Leaves ovate with an abrupt tip, and hairless. Male catkins cylindrical and pendant, to 50 mm long; female catkins ovate, to 15 mm long, and woody when mature; all catkins appear before the leaves. Local in wet habitats.

Order: **BRASSICALES**

RESEDACEAE

Annual or perennial herbs with flowers borne in long spikes; flowers with 4–8 free sepals and petals.

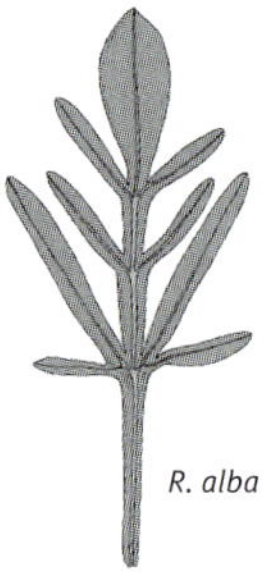

R. alba

Reseda alba **WHITE MIGNONETTE** A medium to tall, erect perennial with stems branched above. Leaves pinnately lobed with 10 or more narrow, untoothed lobes. **Flowers white**, to 9 mm, with 5–6 sepals and petals, the petals all lobed to a third or more of their length. Flowers borne in long spikes. Fruit capsule 4-angled and elliptical, constricted apically and erect when ripe. Common in disturbed areas and sandy waste places.

Reseda lutea **WILD MIGNONETTE** A short to tall leafy perennial, bushy and sometime woody at the base. Leaves stalked and pinnately lobed with 1–4 pairs of leaflets. **Flowers pale greenish-yellow** with 6 petals and sepals, borne in many long narrow spikes. Fruit oblong and erect, 3-parted. Very common in disturbed areas and sandy waste places.

1. *Quercus suber*
2. *Quercus canariensis*
3. *Quercus canariensis*
4. *Quercus lusitanica*
5. *Reseda lutea*

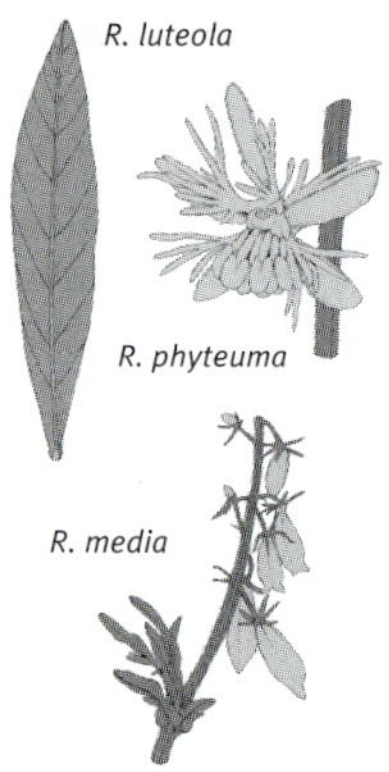

Reseda luteola **WELD** A large, erect biennial to 1.3 m. **Leaves unlobed,** lanceolate and with wavy margins. Flowers greenish-yellow with 4 petals borne in long, slender spikes. Fruit rounded with 3 pointed lobes. Local in disturbed areas and sandy waste places. ***Reseda phyteuma*** is a shorter annual to 50 cm with undivided or slightly lobed leaves (up to 1–2 pairs of lobes in the lower part); flowers white with 6 sepals and petals, and the **fruit distinctly nodding**, with conspicuous and persistent sepals. Local on cultivated ground and roadsides. ***Reseda media*** is very similar to *R. phyteuma* but with many **leaves with up to 8 pairs of lobes**, (leaf segments 8 mm wide or less) and a small terminal lobe, the upper leaves rounded at the base. Dry places.

S. spathulifolia

Sesamoides spathulifolia Similar in form to *Reseda* but a **dwarf perennial or small shrub** with a **lax rosette** from which arise stems to 15 cm, **woody** at the base, and flowers borne in short, dense racemes with bracts; petals 5–6. Coastal slopes and sand dunes; common on Cape St. Vincent, local elsewhere.

BRASSICACEAE

A large family of annual and perennial herbs. Leaves alternate, flowers usually in racemes, very distinctive with 4 sepals and petals forming a cross; stamens 6. Fruit a bipartite capsule. The shape of the ripe fruit is important for identification and can be classified as a silicula (broad and variously shaped) or a siliqua (long and thin).

Isatis tinctoria **WOAD** A large, erect biennial to 2 m. Leaves grey-green, arrow-shaped and clasping the stem above. Flowers yellow in dense and much-branched racemes. **Fruit a pendent**, oblong-elliptical silicula. Sporadic in waste places and on disturbed ground. Used by the Romans as a medicinal plant and cultivated in the Middle Ages as a source of the blue dye indigotine.

Capsella bursa-pastoris **SHEPHERD'S PURSE** A distinctive, sparsely hairy annual with variable leaf shape. Scentless white flowers borne in a long raceme; petals twice as long as the sepals. Silicula longer than wide with straight or convex lateral margins; **heart-shaped**. A very common weed in dry waste places throughout.

Sisymbrium officinale **HEDGE MUSTARD** A bushy, more or less hairless annual or biennial to 90 cm with grey-green dense, ovate, divided leaves with a large terminal lobe. Flowers yellow, small (petals to 4 mm), borne on slender, branched, lax **inflorescences without bracts**. Siliqua to 20 mm, cylindrical, straight, **closely and densely appressed to slender stems**. Common on sandy roadsides.

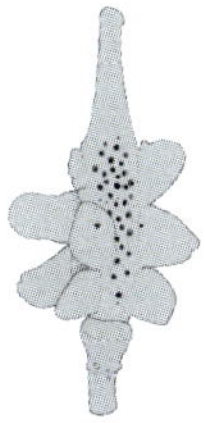
B. erucago

Bunias erucago **SOUTHERN WARTY CABBAGE** A bristly hairy annual or biennial to 60 cm. Lower leaves pinnately lobed or wavy-margined, upper leaves unlobed, toothed or not. Flowers yellow, 8–10 mm across with notched petals. Silicula square in section, with toothed wings on the angles. In waste places and cultivated land.

M. littorea

Malcolmia littorea A densely white-hairy, perennial herb to 40 cm tall, woody at the base and with numerous non-flowering shoots. **Leaves more or less stalkless**, lobed or not. Flowers pink-purple, to 20 mm across. Siliqua long to 65 mm and not beaded. Very common in sandy and rocky maritime habitats. ***Malcolmia lacera*** is similar to *M. littorea* but with **shortly petiolate basal leaves**, smaller flowers, and a siliqua to only 40 mm long. ***Malcolmia ramosissima*** is similar to both species but with a **downy fruit** to 35 mm long.

1. *Reseda luteola*
2. *Sesamoides spathulifolia*
3. *Capsella bursa-pastoris*
4. *Isatis tinctoria*
5. *Sisymbrium officinale*
6. *Malcolmia littorea*

M. incana

Matthiola incana HOARY STOCK A stout, bushy perennial to 40 cm with a woody stock and numerous non-flowering shoots, grey-green and hairy. Leaves lanceolate, usually untoothed and unlobed. Flowers pink, purple or white, the petals 20–30 mm long. Fruit long and thin, to 16 cm, laterally compressed; hairy but not glandular. On sea cliffs and other coastal habitats. **_Matthiola sinuata_** is similar to *M. incana* but smaller and less bushy, the lower leaves deeply pinnately lobed and wavy-margined, the upper leaves entire, flowers pale purple, and the **fruit glandular and sticky** without conspicuous apical horns. **_Matthiola parviflora_** is similar, with deeply pinnately lobed lower leaves, but small flowers with **brownish-purple petals to 10 mm long** and a **cylindrical (not laterally compressed) fruit** to 70 mm long that is **slightly constricted at intervals**.

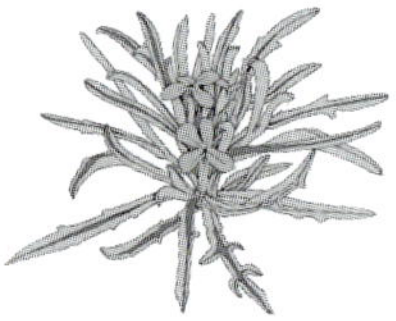

M. sinuata

Lobularia maritima SWEET ALISON A greyish, downy, spreading woody-based perennial with stems to 40 cm, often prostrate. Leaves narrow, untoothed, linear-lanceolate and pointed. Flowers white and sweetly scented, in dense rounded racemes that lengthen in fruit. Silicula flattened, oval or rounded, often hairy, to 3.5 mm long with usually 2 seeds. Very common in sunny coastal environments.

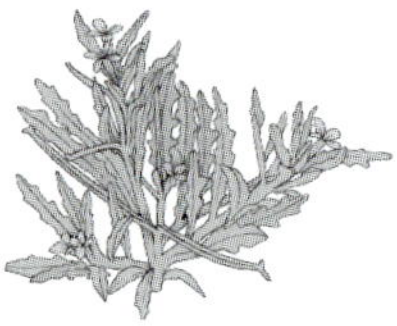

M. parviflora

Ionopsidium acaule A small, easily overlooked tuft-like, virtually stemless annual with a rosette of rounded-ovate, entire to 3-lobed leaves. Flowers solitary, borne on long pedicels in the axils of the leaves. Petals white to lilac, 2–3 times as long as the calyx. Silicula ovoid. A very rare Portuguese endemic; only in the Cape St. Vincent area.

I. umbellata

Iberis umbellata CANDYTUFT A robust, **hairless annual** to 70 cm with linear-lanceolate leaves. Inflorescence a dense, rounded, flat-topped cluster. Petals pink to purple, unequal: 2 long and 2 short. Silicula to 10 mm long, ovate and broadly winged from the base. On rocky slopes and cultivated places; infrequent.

Iberis sampaiana A short, **procumbent**, **bristly hairy annual** to 20 cm high. Leaves spatula-shaped, **pinnatifid** with 3–4 pairs of segments. Inflorescence a dense, rounded, flat-topped cluster; petals white, the outermost strongly radiate. Silicula ovate-rectangular, not distinctly hairy. Very local, on calcareous soils.

Iberis procumbens A short, compact and **bushy, puberulent perennial** with a woody stock from which both non-flowering and flowering stems arise. Leaves broadly spathulate, fleshy and entire or with 1–2 pairs of teeth near the apex. Inflorescence a dense, rounded, flat-topped cluster, **contracted in fruit**. Petals white-lilac. Silicula ovate, broadly winged from the base. On sea cliffs in the Cape St. Vincent area.

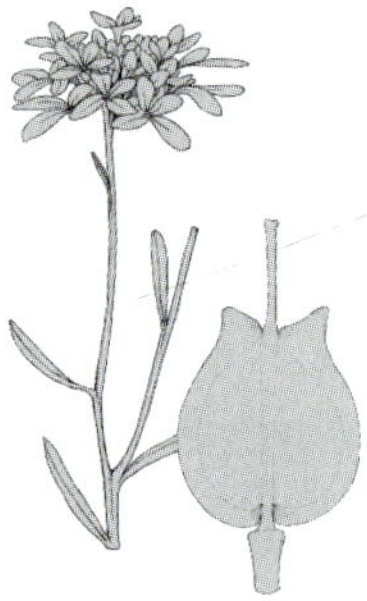

I. linifolia

Iberis linifolia A short, **laxly tuft-like perennial** emerging from a woody stock with numerous erect and puberulent stems to 30 cm. **Leaves linear**, up to 30 mm long, rather fleshy, the lower ones with a few teeth near the apex. Inflorescence a dense, rounded, flat-topped cluster of white flowers. Silicula round-ovate and broadly winged from the base. Local on acid maritime sands.

C. draba

Cardaria draba HOARY CRESS An erect, hairless or slightly downy greyish perennial to 90 cm. Leaves oblong, pointed, toothed and long-stalked at the base, clasping the stem above. Flowers white, to 6 mm across and borne in dense, umbel-like clusters. Silicula heart-shaped with a protruding style, inflated, not splitting. Common in waste places and on roadsides.

Biscutella vicentina An erect perennial to 30 cm with a thick, woody rhizome. Basal leaves numerous, to 60 mm long, and densely clustered. Leaves oblong, wavy at the margin, and minutely hairy. Inflorescence a dense raceme of yellow flowers, elongated in fruit. **Silicula very large, to 11 × 19 mm, with a membranous margin**. A rare endemic, restricted to sandy soils in the Cape St. Vincent region; sometimes treated as a subspecies of *Biscutella sempervirens* which occurs in southern Spain.

1. *Lobularia maritima*
2. *Ionopsidium acaule*
3. *Iberis procumbens*
4. *Biscutella vicentina*

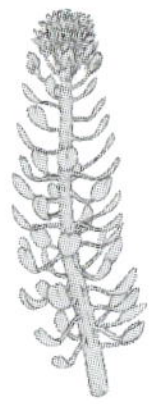

L. heterophyllum

Lepidium heterophyllum **Smith's cress** A perennial herb with a woody stock and numerous ascending stems to 45 cm tall; grey-green with short, spreading hairs. Basal leaves ovate to elliptical, and soon-falling; upper leaves narrowed into short stalks; uppermost leaves stalkless and triangular. Flowers small to 4 mm across, **white with 6 violet anthers**. Fruit smooth and flat and shield-like with a notch at the tip. Disturbed grassy and sandy habitats.

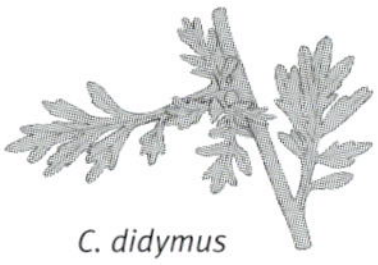

C. didymus

Coronopus didymus **lesser swinecress** A small, spreading or ascending biennial up to 30 cm. Leaves divided, almost feathery and carrot-like, first in a rosette, then scrambling. Flowers insignificant, petals 0.5 mm, shorter than the sepals, or absent. Racemes elongated in fruit, pedicels longer than the silicula. Silicula 1.5 x 3 mm, veined and with a notch at the apex, style absent. Common in sandy waste places. ***Coronopus squamatus*** is similar, but with flowers with white petals to 2.5 mm across, and ridged, kidney-shaped (not dumb-bell-shaped) fruits.

1

2

3

4

5

Diplotaxis vicentina An annual with more or less hairless, **leafless stems** and deeply divided leaves at the base; the segments unequal (the terminal larger than the laterals), with short hairs on both surfaces. Flowers sulphur-yellow. Siliqua elongated to 45 mm with globose seeds. A rare endemic restricted to sandy soils in the Cape St. Vincent region; sometimes treated as a subspecies of *Diplotaxis siifolia*, which occurs throughout the southern Iberian Peninsula.

Brassica nigra **BLACK MUSTARD** An annual to 1 m with stems branched from the middle or from near the base. Lower leaves round to lyre-shaped, and pinnatisect (lobed to the midrib), with 1–3 pairs of lateral lobes and a much larger terminal lobe; **bristly on both surfaces**. Upper leaves linear-oblong, usually entire and without bristles. **Leaves petiolate**. Petals yellow. Siliqua to 20 mm long and slender (more so gradually towards the seedless beak), on short pedicles appressed to the stem. Local on maritime grassy and sandy places.

B. barrelieri

Brassica barrelieri An annual to 50 cm. Lower leaves numerous and clustered in a rosette, very shortly petiolate, pinnatifid with 7–10 pairs of lanceolate lobes; **bristly,** especially along the rachis. Stem leaves few, **sessile**, entire and hairless. Petals conspicuous, yellow with red veins or whitish. Siliqua to 60 mm long and constricted at intervals. Local on sandy maritime soils.

S. alba

Sinapis alba **WHITE MUSTARD** A tall, normally bristly annual (sometimes hairless). **Leaves all stalked** and pinnately lobed. Flowers pale yellow. Siliqua to 40 mm long, the beak flattened and sword-like, as long as, or exceeding the valves. Common on disturbed, cultivated and waste areas. The source of white mustard. ***Sinapis arvensis*** is similar but with lanceolate **unstalked upper leaves**, and a larger siliqua to 45 mm long, with a **conical beak shorter than the 3–7 valves**. Habitats similar to those of *S. alba*.

Hirschfeldia incana **HOARY MUSTARD** A tall, lax, erect annual to 1 m. Lower leaves stalked and pinnate with an oblong end-lobe and up to 9 pairs of lateral lobes, the uppermost unlobed. Flowers pale yellow, borne in crowded terminal racemes, superficially similar to *Sisymbrium* spp. Fruit closely appressed to the stem, peg-shaped with a swollen, single-seeded upper segment, and flattened, 2–6-seeded lower segment. Disturbed, sandy ground; frequent.

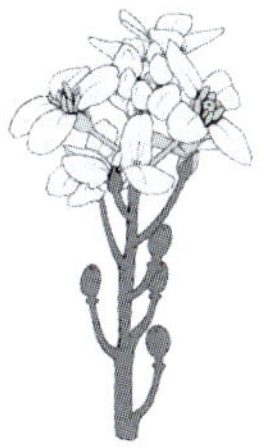

C. maritima subsp. *aegyptica*

Cakile maritima* subsp. *aegyptiaca **SEA ROCKET** A variable, short, rather succulent, spreading, hairless annual. Leaves grey-green, irregularly pinnately lobed, the lobes narrow and untoothed, or undivided. Flowers lilac to white, the petals to 15 mm long, borne in dense racemes that elongate significantly when in fruit. Siliqua brown and succulent; bipartite, **the lower segment with an arrow-shaped base**, the upper ovate and 4-angled. Common on maritime sands throughout.

1. *Coronopus didymus*
2. *Coronopus squamatus*
3. *Diplotaxis vicentina*
4. *Diplotaxis vicentina*
5. *Cakile maritima* subsp. *aegyptiaca*

R. rugosum

Rapistrum rugosum An annual to 60 cm, bristly hairy below and hairless above. Lower leaves pinnate, often toothed and petiolate. Petals lemon yellow, to 10 mm long. Silicula 3–10 mm, the upper segment ovoid-globose and **abruptly contracted into the beak**, the lower segment cylindrical. Common in ruderal maritime habitats.

Crambe hispanica A slender, bristly hairy annual to 1 m. Lower leaves rather shiny, round-lyre-shaped, divided and with a single large, kidney-shaped terminal lobe, and 0–2 pairs of insignificant lateral lobes. Petals white, to 4 mm, sometimes purple at the base borne in a **much-branched panicle**. Silicula 2–4 mm with a short lower part and a **globose upper part containing a single seed**. A range of habitats including arable land and coastal *matos*.

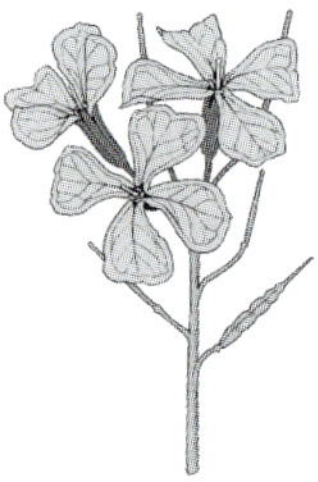
R. raphanistrum

Raphanus raphanistrum **WILD RADISH** A variable, short to tall, bristly annual to 1.5 m, erect and branched. Flowers white to pale yellow, **with lilac or reddish veins**, borne in branched racemes. Siliqua to 90 mm long, **jointed and beaded**. Very common on arable land, sea cliffs, sandy and waste places throughout. ***Eruca sativa*** is a superficially similar bristly annual to 1 m with petiolate, lobed leaves with a large terminal lobe, erect, purple sepals and flowers white or pale yellow and purple-veined, but differs in having a smaller, **unbeaded siliqua to 25 mm long with a flattened, sword-shaped beak**. Habitats similar to those of *R. raphanistrum*. ***Erucastrum gallicum*** is a hairy annual with lobed leaves. Flowers pale yellow, to 15 mm across, the stalk with **pinnately lobed bracts**. Disturbed, sandy habitats.

Teesdalia nudicaulis An annual to 25 cm with few or no stem leaves and a basal rosette of pinnately lobed leaves. Flowers to 2 mm across with white petals, **2 of which are much shorter than the other pair**. Fruit heart-shaped and notched above, to 4 mm. Local; frequent on Picota.

Brassica oleracea **CABBAGE** A tall, 'tree-like' cabbage to 1 m tall or more with leaves normally stripped from the base upwards (for culinary purposes) and sprays of white flowers. Commonly cultivated in gardens and along roadsides for making vegetable soup according to a local recipe.

Order: MALVALES

MALVACEAE

A large and taxonomically complex family of herbs or shrubs with star-shaped (stellate) hairs, and with stipules. Leaves alternate and usually palmately lobed. Flowers conspicuous, actinomorphic and cosexual with 5 sepals and petals; stamens many. Fruit a nut or capsule.

M. trifida

Malope trifida A short to medium, virtually hairless annual with a single, erect stem to 1.5 m. Leaves long-stalked, wavy-edged and rounded below, with 3–5 triangular lobes above. Flowers deep purple-red, the petals 30–60 mm long, not notched. Cultivated in gardens and perhaps naturalised.

1. *Raphanus raphanistrum*
2. *Brassica oleracea*
3. *Malva sylvestris*

M. sylvestris

M. neglecta

M. nicaeensis

M. hispanica

Malva sylvestris **COMMON MALLOW** A variable, erect to spreading perennial herb to 1.5 m. Leaves roughly kidney-shaped or heart-shaped with 3–7 toothed lobes. Flowers borne in clusters of 2 or more, pink or purple **with darker veins**, the petals to 30 mm long and bearded; **petals about 4 times the length of the downy sepals**. **Epicalyx** (bracts below the true calyx) **segments free to the base**. Fruit sharply angled, hairy or not, and netted. Common on disturbed ground; frequently confused with *Lavatera cretica*, which has epicalyx segments **fused at the base**, and less strongly marked flowers. ***Malva neglecta*** is similar to *M. sylvestris* but a smaller annual to 60 cm with leaves less deeply lobed, **petals 2–3 times the length of the sepals**, and the fruit smooth. Habitats similar to those of *M. sylvestris*.

Malva nicaeensis is similar to *M. sylvestris* but with semicircular leaves without a heart-shaped base, sepals hairless beneath, and pale lilac **petals with less prominent darker veins**. Habitats similar to those of *M. sylvestris*.

Malva hispanica An erect annual to 70 cm with long hairs. Leaves small, to 30 mm wide, semicircular and **scarcely lobed or unlobed**. Flowers normally solitary, borne in the leaf axils, epicalyx lobes linear to triangular, sepals diamond-shaped to ovate, petals about twice the length of the sepals, pale pink throughout. Fruit hairless and without ridges. Sandy and dry habitats.

1

2

3

M. tournefortiana

Malva tournefortiana A sparsely hairy, erect perennial with **palmate leaves deeply dissected into narrow, linear segments**. Flowers solitary in the leaf axils, bright pink, the petals about twice as long as the sepals. Epicalyx lobes 3, linear and hairless or sparsely hairy. Flower-stalks with star-like (not simple) hairs. Fruit angled and hairless. On cultivated land and dry, fertile places.

M. parviflora

Malva parviflora A short to medium, hairy or hairless annual. Leaves rounded to heart-shaped with 3–7 shallow, rounded and toothed lobes. Flowers borne in clusters of 2–4, pale mauve or lilac, **small** (the petals less than 5 mm long), **short-stalked** with linear to lanceolate epicalyx lobes. **Fruit with enlarged, spreading sepals**, fruit strongly netted and hairy or not.

L. cretica

Lavatera cretica **SMALL TREE MALLOW** An erect or ascending annual or biennial to 1.5 m, densely or sparsely covered in star-shaped hairs. Lower leaves rounded to heart-shaped, shallowly 3–7-lobed, the upper leaves deeply 5-lobed. Flowers pink, borne in the axils of the leaves in clusters of 2–8, on unequal stalks shorter than the leaves. Petals to 20 mm long, epicalyx segments to 6 mm long, and **free almost to the base (but there fused),** shorter than the long-pointed sepals; sepals slightly enlarged in fruit. Fruit smooth or slightly ridged. Common on waste ground. Often confused with *Malva sylvestris*, which is similar but with both **simple and star-shaped hairs** on the flower stalks, and narrow epicalyx segments that are **free completely to the base (not fused)**. ***Lavatera mauritanica*** is similar to *L. cretica* but with **sepals much enlarged in fruit**, and markedly ridged fruits.

L. mauritanica

L. arborea

Lavatera arborea **TREE MALLOW** A robust, **woody** biennial to 3 m tall, downy above with star-shaped hairs. Leaves large, rounded and palmate with 5–7 lobes, up to 20 cm long and velvety. Flowers borne in the leaf axils in clusters of 2–7, forming a long, terminal inflorescence. Flowers pink-purple with darker veins. Epicalyx segments to 10 mm long, **exceeding the sepals** and greatly enlarged in fruit and fused at the base. Fruit hairy or not, and sharply angled. Common on maritime rocks. ***Lavatera maritima*** is rather similar to *L. arborea* but with **white-felted younger parts**, pale bluish-pink flowers **solitary or in pairs** with distinct gaps between the petal bases. Habitats similar to those of *L. arborea*.

L. maritima

Lavatera olbia TREE LAVATERA A robust, erect, laxly branched **shrub to 2 m** with young parts covered in star-shaped hairs. Lower leaves 3–5-lobed, the upper leaves only shallowly lobed. Flowers solitary in the axils of the leaves, forming hollyhock-like spikes; petals purple, the epicalyx segments slightly shorter than the sepals, both ovate and pointed. Fruit hairy or bristly, not ridged. Various habitats.

L. trimestris

Lavatera trimestris ANNUAL LAVATERA An erect, somewhat **bristly** annual to 1.2 m, with long-stalked, rounded to heart-shaped, toothed leaves; the upper shallowly 3–7-lobed, deep shiny green. Flowers large, the petals 25–40 mm long, pink (sometimes white), solitary. Epicalyx segments shorter than the sepals; sepals fused over most of their length, and **enlarging in fruit**. Fruits hairless, ridged, and **covered by a disc-like expansion** projecting from the central axis. On cultivated, fallow and waste ground.

L. triloba

Lavatera triloba* subsp. *triloba MUSK-SCENTED LAVATERA A **glandular-hairy, musk-scented**, greyish or whitish hairy perennial, woody at the base. Leaves rounded to heart-shaped, long-stalked, slightly 3-lobed, flat or wavy-margined. **Flowers borne in clusters** of 3–7, purple, sometimes flushed with yellow, or yellow throughout, the petals to 30 mm long. Epicalyx lobes shorter than the sepals; sepals enlarged in fruit. Fruit hairless or glandular-hairy. Normally damp or saline habitats.

A. hirsuta

Althaea hirsuta ROUGH MARSH MALLOW A short to medium, slender annual to 60 cm with simple and star-like hairs with swollen bases. Lower leaves rounded and kidney-shaped, to 60 mm across, long-stalked and toothed or shallowly lobed, more deeply palmately lobed further up the stem. Flowers pink, solitary, long-stalked and small, cup-shaped, the petals scarcely exceeding the sepals. Epicalyx segments 6–9, and nearly as long as the sepals. Fruit hairy and ridged. Dry fallow land and maritime waste places.

Althaea officinalis MARSH MALLOW A tall, velvety-downy perennial to 2 m with all hairs star-shaped. Leaves triangular-ovate, entire or scarcely 3–5-lobed, often folded and fan-like, to 90 mm across. Flowers solitary or clustered in the leaf axils, pale lilac-pink, borne on stalks shorter than their adjacent leaves. Fruits hairy, sepals curved outwards in fruit. In damp and saline coastal habitats.

H. syriacus

Hibiscus syriacus COMMON HIBISCUS An erect shrub to 3 m with diamond-shaped, toothed leaves that are somewhat 3-lobed and short-stalked. Flowers variable in colour from pink to white with a purplish centre, solitary or in pairs in the upper leaf axils. Capsule yellow-hairy. Widely planted as an ornamental. ***Hibiscus rosa-sinensis*** has shiny, ovate, sometimes slightly irregularly toothed leaves, and conspicuous and showy bright red flowers. Very commonly planted in towns and gardens.

Hibiscus esculentus OKRA A rather large perennial with broad, palmately-lobed leaves to 20 cm across, held upright on long stems. Flowers cream with a dark purple centre, to 80 mm across with 5 petals. Fruit an edible pod to 18 cm long. Occasionally cultivated in gardens in the Algarve.

---

1. *Lavatera cretica*
2. *Lavatera arborea*
3. *Hibiscus rosa-sinensis*

Dombeya × cayeuxii An exotic, probably of hybrid origin of *Dombeya burgessiae* and *Dombeya wallichii*, which grows to 5 m tall with large, drooping leaves and pendent inflorescences of pink flowers that are long-persisting. Commonly planted in towns across the Algarve.

CYTINACEAE

A family of obligate, endophytic parasites of the roots of Cistaceae (in the Mediterranean at least). Previously considered to be related to the Rafflesiaceae, but recently allied with the Malvales. Plants characterised by their leafless, fleshy inflorescences that lack chlorophyll. Unisexual, ant-pollinated flowers in a dense spike, the lower female, the upper male. Seeds numerous in a viscid pulp, dust-like and windborne when dry. Distinct genetic races appear to be defined by host specificity and are difficult to distinguish by morphology alone.

C. hypocistis

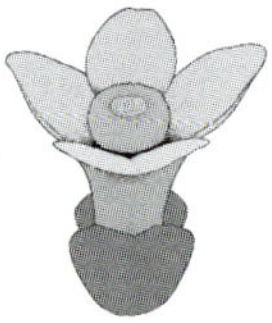

C. hypocistis subsp. *macranthus*

C. hypocistis subsp. *hypocistis*

Cytinus hypocistis **CYTINUS** Stem to 50 mm, usually less. Scale leaves ovate-oblong, yellow, orange or red. Flowers 5–15, subtended by 2 bracteoles, which are the same colour as the leaves; perianth bright yellow. Two subspecies occur in the area and are defined largely by ecology; they are continuously variable and difficult to distinguish, especially when in bud. **Subsp. *macranthus*** has red scale leaves when young **fading orange-yellow or yellow when flowers mature**, and a **perianth that exceeds the bracteoles** at anthesis. Often forms large clumps, normally parasitic on *Halimium halimifolium* (also on *Cistus palhinae*) in dunes and other sandy habitats. Common on fixed coastal dunes. **Subsp. *hypocistis*** has bright red scale leaves and bracteoles contrasting the yellow flowers, **remaining red when flowers mature** and a **perianth scarcely exceeding the bracteoles**. Inflorescences solitary or in small clumps. Normally parasitic on *Cistus monspeliensis* and *C. populifolius* on rocky hill slopes and on the *matos*, on clay-based soils. Locally common where suitable host species are frequent.

Cytinus ruber Similar to *C. hypocistis*, but with **crimson (not red-orange) scale leaves** and bracteoles contrasting a **white-pale pink (not yellow) perianth** that slightly exceeds the bracteoles. On pink-flowered *Cistus* spp., particularly on *C. albidus* on dry slopes and *matos*. Throughout the Algarve, though less common than *C. hypocistis*; locally common on coastal *matos* between Burgau and Salema on southern cliff tops.

THYMELAEACEAE

Small shrubs with simple, untoothed leaves. Flowers in clusters or racemes, cosexual, calyx a tube, petal-like; true petals absent; stamens 8, fused to the surface of the calyx-tubes; style solitary. Fruit a berry or nut.

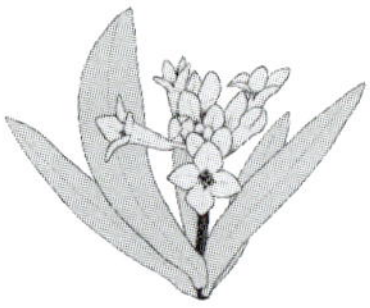

D. gnidium

Daphne gnidium **DAPHNE** An erect, lax, virtually hairless shrub to 2 m with branches bare beneath; superficially rather *Euphorbia*-like when not in flower, but without a milky latex. Leaves pale green, leathery, linear and pointed. Flowers cream-white, to 6 mm across, borne in dense panicles. Berry deep red then black. Common on the *matos* and dry slopes.

Thymelaea hirsuta A dwarf evergreen shrub to 50 cm with white-downy stems; superficially similar to shrubby Amaranthaceae, with small, scale-like succulent leaves overlapping along the stem. Leaves shiny green and

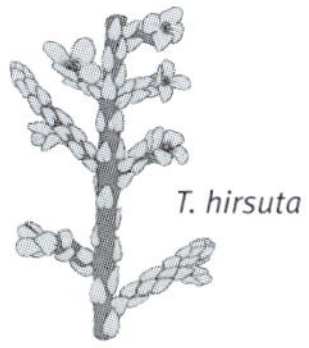

hairless outside, white-downy inside. Flowers unisexual or cosexual; yellowish, borne in inconspicuous clusters of 2–5. Local in dry and sandy places. ***Thymelaea villosa*** is similar to *T. hirsuta* but with leaves equally hairy on both surfaces, dioecious and with solitary flowers. Local, in habitats similar to those of *T. hirsuta*. ***Thymelaea passerina*** is a **hairless annual** with greenish or sometimes white flowers.

1. *Dombeya × cayeuxii*
2. *Cytinus hypocistis* subsp. *macranthus*
3. *Cytinus hypocistis* subsp. *hypocistis*
4. *Cytinus ruber*
5. *Daphne gnidium*

1. *Cistus crispus*
2. *Cistus albidus*
3. *Cistus salvifolius*
4. *Cistus monspeliensis*
5. *Cistus ladanifer* blotched form
6. *Cistus ladanifer*
7. *Cistus palhinhae*
8. *Cistus populifolius*

CISTACEAE

An important family of herbs and shrubs in the Mediterranean. Leaves normally opposite, stipules often present. Flowers often showy, cosexual, solitary or in lax clusters, sepals 3–5, petals 5, stamens numerous. Fruit a capsule.

Cistus crispus A short, compact, aromatic shrub to 40 cm. Leaves oblong to elliptical, to 40 mm long with a distinct **undulate margin**, 3-veined, grey-green, stalkless and hairy. Flowers short-stalked, pink, to 40 mm, borne in few-flowered clusters; sepals 5, hairy. Locally common on *matos* at a range of altitudes.

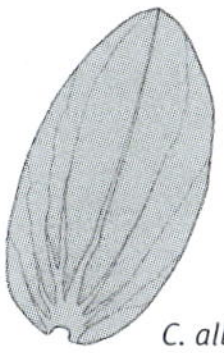

C. albidus

Cistus albidus **GREY-LEAVED ROCK ROSE** A lax or compact bush to 1 m. Leaves oblong to elliptical, fairly large to 25 mm long, **grey-white** with soft hairs. Flowers on stalks to 20 mm or shorter, pink, to 60 mm across borne in few-flowered clusters; sepals 5, hairy. Very common, particularly at low altitude, often subdominant on the *matos* throughout the region; often growing in large patches.

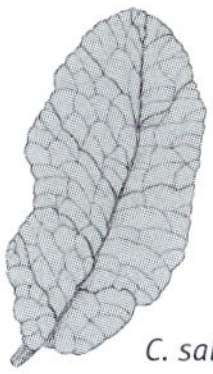

C. salvifolius

Cistus salvifolius **SAGE-LEAVED ROCK ROSE** A low, spreading, almost prostrate shrub to 30 cm. **Leaves sage-like**, short-stalked, mid green, ovate-elliptical with a rounded or wedge-shaped base; to 40 mm long, 3-veined and hairy on both surfaces. Flowers white, to 50 mm across, solitary or in clusters of up to 4; sepals 5, the outer 2 with a heart-shaped base. Very common on *matos* and fixed dunes throughout the region.

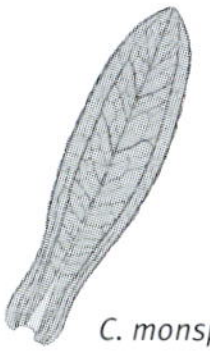

C. monspeliensis

Cistus monspeliensis **NARROW-LEAVED ROCK ROSE** A **slightly sticky bush** to 1 m, lax below, compact above. **Leaves narrow**, linear, scarcely tapered at the base and unstalked, mid green. Flowers small, to 30 mm across or smaller, white; 5 sepals, the outer 2 wedge-shaped at the base. Very common on the *matos* throughout the region, often growing in small patches. ***Cistus psilosepalus*** is similar to *C. monspeliensis* but with oblong leaves and **larger flowers** 40–60 mm across borne in more or less symmetrical cymes of 1–5; sepals 5, the outer 2 **heart-shaped** at the base. Habitats similar to those of *C. monspeliensis*, but local.

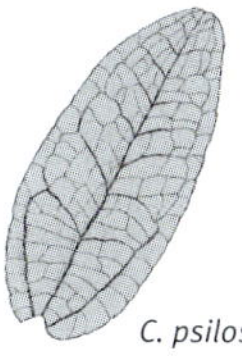

C. psilosepalus

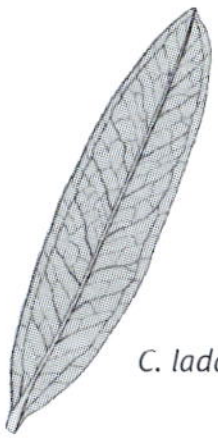

C. ladanifer

Cistus ladanifer **GUM ROCK ROSE** An aromatic, **extremely sticky**, lax, erect shrub to 2 m. Leaves linear-lanceolate, dark green (paler beneath), to 80 mm long, 3-veined in the lower third, and scarcely stalked. Flowers solitary, **large**, to 10 cm across, white, often with a crimson blotch at the base of each petal; sepals 3. Very common, often a dominant component of the *matos* throughout the region. ***Cistus palhinhae*** is very similar to (and possibly an ecotype of) *C. ladanifer*, but much **shorter** (to 50 cm), with leaves ovate to wedge-shaped and broadest above the middle; to 60 mm long. Flowers **unspotted** and white. A rare endemic of the Cape St. Vincent area; locally abundant on maritime cliffs on Cape St. Vincent. Populations are continuously variable with *C. ladanifer* and the extent to which morphological traits unique to this species are a result of its distinct ecology (wind-swept cliffs) has not been established.

C. populifolius

Cistus populifolius A spreading, much-branched shrub to 1.5 m. **Leaves stalked**, ovate with a **heart-shaped base**, to 10 cm long, hairless and smooth with an undulate margin. Flowers white, to 60 mm across borne in clusters of up to 6; sepals 5, the outer 2 with a heart-shaped base, becoming red when the petals fall. Locally common on the *matos* inland.

Halimium Small, often greyish shrubs like *Cistus* but with 3 (not 5 or 10) valves to the capsule, sepals 3 or 5, the outer 2 much smaller, and flowers always yellow or white.

H. halimifolium

Halimium halimifolium A dense, silvery-grey shrub to 1 m. Leaves elliptical to oblong, broadest above the middle; to 40 mm long, and downy when young, with silvery scales and star-like hairs when mature. Flowers yellow, often with a black spot at the base of the petals; to 30 mm across, borne in terminal clusters; **sepals 5**. Very common to subdominant on fixed coastal dunes and sandy slopes. **Subsp. *multiflorum*** produces denser panicles of flowers; sepals with numerous star-like hairs, and petals narrow and heart-shaped at the base, and broader above.

H. atriplicifolium

Halimium atriplicifolium Similar in form to *H. halimifolium* but taller, to 1.5 m. Leaves of non-flowering shoots 3-veined and short-stalked, those of the flowering shoots pinnately veined with a heart-shaped base and unveined; all with silvery scales and star-like hairs. Flowers yellow, with petals spotted at the base; to 50 mm across, with 3 hairy sepals, borne on **densely hairy, leafless branches**. Very common in maritime habitats, though in smaller numbers than *H. halimifolium*.

H. lasianthum

Halimium lasianthum A compact, grey, densely hairy shrub to 1 m with oblong leaves to 40 mm long that are dark green above, woolly white below, and **pointed**. Flower stalks and sepals with long, **silky hairs**, the sepals often with **purple bristles**. Flowers to 30 cm across, yellow and unspotted or with a brown spot at the base of each petal. Local on sandy slopes and cork oak forests. **Subsp. *formosum*** has larger flowers to 60 mm across with petals spotted well above the base.

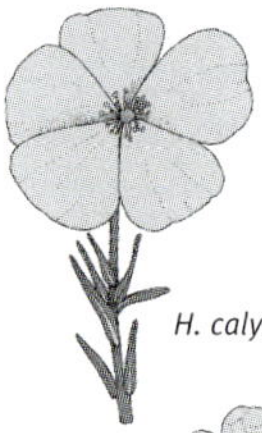

H. calycinum

Halimium calycinum YELLOW ROCK ROSE A low-growing, almost mat-forming shrub to 50 cm. **Leaves linear, with down-turned margins**, somewhat rosemary-like, small, to 35 mm long and less than 4 mm wide; shiny and hairless above, white-hairy below. Flowers solitary or in clusters of 2–5; bright yellow, to 15 mm across; sepals hairless. Very common and often abundant to sub-dominant on coastal sands, and locally common on *matos* at a range of altitudes.

H. verticillatum

Halimium verticillatum A low-growing shrub to 50 cm, rather similar to *H. calycinum* but with sparse star-like hairs. Leaves with down-turned margins or not. **Flowers white**, borne in clusters of 4–5 on very thin flower stalks; sepals hairy to woolly. In oak and pine woods; local.

Tuberaria Annuals or perennials with basal rosettes and erect flowering stems. Flowers yellow, sepals 5, the outer 2 smaller; capsule 3-valved.

T. guttata

Tuberaria guttata ANNUAL ROCK ROSE A very variable, hairy, low annual with a basal leaf rosette that dies when mature, and a normally unbranched flowering stem. Leaves elliptical to ovate, often with a down-turned margin. Flowers yellow, to 20 mm across with or without a dark brown or purple spot at the base of each petal; flower stalks longer than the sepals at the point of flowering. Common in sandy habitats and waste places.

1. *Halimium halimifolium*
2. *Halimium atriplicifolium*
3. *Halimium atriplicifolium*
4. *Halimium calycinum*
5. *Tuberaria guttata*
6. *Tuberaria guttata* unspotted flower form

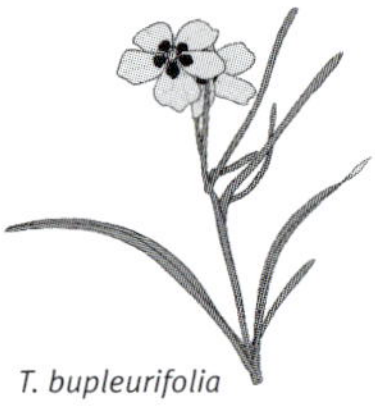

T. bupleurifolia

Tuberaria bupleurifolia is similar but with narrow leaves with strongly down-turned margins, and flowers to just 13 mm across; coastal sands.

Tuberaria lignosa A short perennial with a branched, woody stock and with persistent plantain-like leaf-rosettes from which flower stems to 30 cm arise. Leaves ovate to elliptical, gradually tapered towards the base, 3-veined and white-hairy beneath. Flowers yellow and unspotted. Local in scrubby and sandy habitats.

T. lignosa

Tuberaria major is similar to *T. lignosa* but with shiny, ovate-elliptical, 3-veined **leaves that are distinctly stalked**. Inflorescences borne on long stalks; flowers with 5 bright yellow **petals, each with a prominent brown-red dark spot at the base**. A very rare endemic; on coastal sands between Albufeira and Olhão, particularly to the west of the Ria Formosa, in Pontal, Ludo and Campo de Gambelas; frequently in *Stauracanthus*-dominated vegetation.

Helianthemum Dwarf shrubs or herbs with opposite leaves and flowers borne in 1-sided clusters. Sepals 5, the outer 2 smaller; stamens numerous, style long and S-shaped. Fruit a 3-valved capsule.

H. nummularium

Helianthemum nummularium COMMON ROCK ROSE A very variable prostrate to ascending dwarf shrub. Leaves oblong to lanceolate with **flat or almost flat margins**, hairy or hairless above, and grey, or white-hairy beneath. Stipules small and **leaf-like**. Flowers yellow, to 20 mm, solitary or in clusters of up to 12; sepals hairless. Dry grassy and scrubby habitats; local.

1. *Tuberaria major*
2. *Helianthemum organifolium*
3. *Fumana thymifolia*

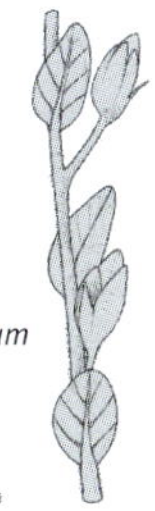
H. salicifolium

Helianthemum salicifolium WILLOW-LEAVED ROCK ROSE A low, hairy, branched, erect or spreading **annual** to 30 cm. Leaves ovate to lanceolate, flat and short-stalked. Flowers yellow, to 22 mm, borne in lax clusters; petals narrow, equalling the sepals; bracts large and leafy; **flower stalks spreading in fruit**, upturned at the apex. On very dry sandy and rocky slopes. **Helianthemum ledifolium** is similar to *H. salicifolium* but taller to 60 cm, **hairier** and with **petals shorter than the sepals**, and **flower stalks erect in fruit**. Habitats similar to those of *H. salicifolium*. **Helianthemum sanguineum** is similar to *H. salicifolium* but very small (usually less than 10 cm), **sticky**, and with **flower stalks strongly bent downwards in fruit**.

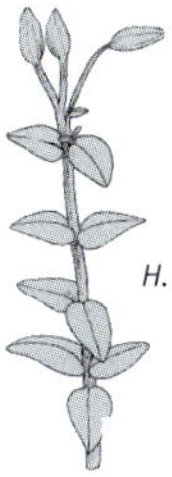
H. organifolium

Helianthemum organifolium A low, spreading, prostrate or ascending dwarf shrub. **Leaves ovate to rounded** and **heart-shaped at the base**, green on both surfaces with star-shaped hairs beneath, and leafy stipules. Flowers yellow, to 8 mm borne in branched inflorescences. Common in rocky and sandy calcareous habitats. **Helianthemum marifolium** is similar but with **leaves white-woolly beneath**, and **leaves without stipules**. Stony and rocky habitats.

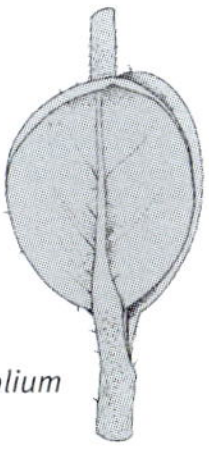
H. marifolium

Helianthemum pilosum A low, ascending dwarf shrub to 30 cm with **white-hairy branches**. Leaves longer further up the stem, to 20 mm, more or less **linear**, green above and grey beneath with down-turned margins. **Flowers white** with yellow stamens; sepals with few hairs or hairless. On dry clay soils.

Helianthemum villosum A small, slender, erect **annual** to 20 cm with grey-green lanceolate leaves to 8 mm with star-like hairs on both surfaces. Stipules linear and to almost half as long as the leaves. Flowers borne in dense inflorescences, more or less stalkless, yellow; the narrow petals inconspicuous and shorter than the woolly sepals. Dry scrub and clay soils.

Fumana Dwarf shrubs similar to *Helianthemum* with narrow, lanceolate to linear, usually alternate leaves. Outer stamens sterile.

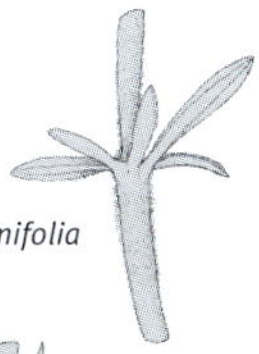
F. thymifolia

Fumana thymifolia THYME-LEAVED FUMANA A small, much-branched somewhat thyme-like (but scentless) dwarf shrub, to 20 cm, with erect or ascending branches. **Leaves opposite** at least below, unequal on the stem and reduced above, linear to elliptical, hairless or stickily hairy with a down-turned margin. Stipules present, as well as short shoots in the leaf axils. Flowers yellow, to 14 mm borne in 3–9 flowered clusters; the inner sepals much larger than the outer 2, membranous and green-veined. **Fumana laevipes** is similar to *F. thymifolia* but with alternate, linear, **bristle-like leaves**. Habitats similar to those of *F. thymifolia*.

F. laevipes

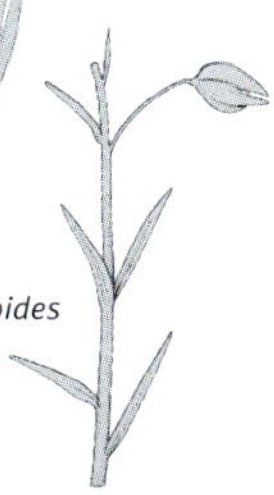
F. ericoides

Fumana ericoides An erect, short dwarf shrub to 20 cm. Leaves alternate, linear, hairless or stickily hairy and evenly spaced along the stem and not reduced above. Flowers yellow, to 16 mm borne on **stalks much longer than the adjacent leaves**, and spreading and down-turned at the apex in fruit. Dry slopes and *matos*.

Order: **SAPINDALES**

ANACARDIACEAE

A large, mostly tropical family of shrubs, trees and lianas. Leaves normally pinnate, flowers 5-parted with 5–10 stamens, fruit a drupe.

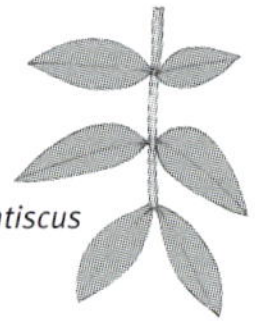
P. lentiscus

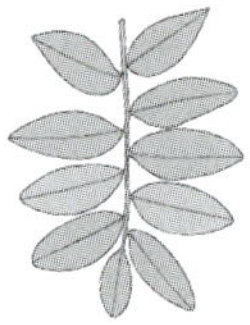
P. terebinthus

Pistacia lentiscus **MASTIC** A small, evergreen, dark green tree or shrub to 8 m. Leaves dark green, **pinnate without an end leaflet**; leaflets 6–18, oval, leathery and untoothed, borne on winged stalks. Individual flowers rather inconspicuous, borne in dense **spike-like clusters**, the male with dark red anthers, the female greenish. Fruit spherical, red then black and shiny. Common to abundant, forming a principle component of the *matos*. Mastic (the resin) was formerly used as a chewing gum. ***Pistacia terebinthus*** is a similar deciduous tree with **pinnate leaves with an end leaflet**, and flowers borne in long-branched panicles. Less common in the area than *P. lentiscus*, but in similar habitats on rocky slopes. ***Pistacia raportae*** is possibly a hybrid of the previous two species, with vegetative parts similar to those of *P. lentiscus* but with female flowers like *P. terebinthus*; fruit whitish. Distribution improperly known in the Algarve.

S. molle

Schinus molle **CALIFORNIAN PEPPER TREE** A small evergreen tree to 10 m with slender, **hanging branches**. Leaves pinnate with 7–13 pairs of linear-lanceolate, toothed leaflets that are hairy when young. Flowers yellow-white, borne in small, much-branched, hanging inflorescences; sepals and petals 5. Fruit pink, spherical. Planted as an ornamental; rarely naturalised.

SIMAROUBACEAE

A small, largely tropical family of trees or shrubs with bitter-tasting bark.

Ailanthus altissima **TREE OF HEAVEN** A large dioecious, deciduous tree to 20 m, suckering freely, with grey bark. Leaves alternate, pinnate, more or less hairless with up to 25 lanceolate to ovate leaflets with few teeth near the base, each tooth with a gland beneath. Flowers green-cream, borne in large terminal panicles, mostly 5-parted. Widely planted as an ornamental.

MELIACEAE

A family of mostly trees and shrubs characterised by alternate, usually pinnate leaves without stipules, and by bisexual (or cryptically unisexual) flowers borne in panicles, cymes, spikes or clusters.

Melia azedarach **INDIAN BEAD TREE** A deciduous tree to 15 m tall with furrowed bark. Leaves alternate and twice pinnate; leaflets elliptical and toothed or lobed. Flowers lilac and scented, borne in panicles. Fruit a berry to 14 mm across, yellow and **long-persisting**, even when the tree sheds its leaves. Native to India and China but widely planted along roadsides in the Algarve.

---

1. *Pistacia lentiscus*
2. *Pistacia lentiscus* fruits
3. *Pistacia terebinthus*
4. *Melia azedarach*
5. *Ruta angustifolia*

RUTACEAE

A large and widely distributed family of herbs shrubs and trees. Leaves with translucent glands. Flowers normally 4–5-parted. Fruit a berry, capsule or drupe; the whole plant very aromatic.

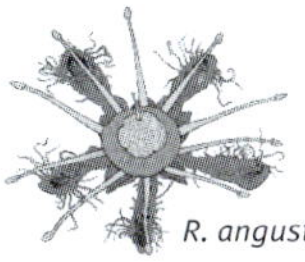

R. angustifolia

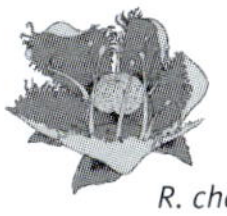

R. chalepensis

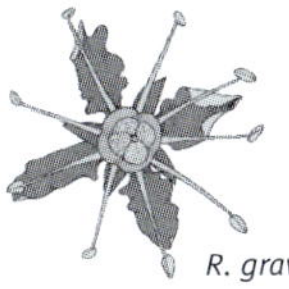

R. graveolens

Ruta angustifolia A hairless, light grey-green, woody-based and densely branched subshrub to 45 cm. Leaves alternate, 2–3-times pinnately lobed, blue-green and very aromatic when crushed, short-stalked below, stalkless above. **Lower bracts scarcely wider than their branches.** Flowers yellow, borne in a loose, **glandular-hairy**, branched cluster; petals widely separated and **with finely toothed margins**, the fringe as long as the width of the petal. Common in dry, rocky habitats. ***Ruta chalepensis*** is similar to *R. angustifolia* but with **lower bracts much wider than their branches**, a **hairless inflorescence**, and petal fringes shorter than the width of the petal. Habitats similar to those of *R. angustifolia*. ***Ruta graveolens*** is again similar to *R. angustifolia* but distinctly grey-blue and with denser branches, and **toothed petals**. Habitats similar to those of *R. angustifolia*. ***Ruta montana*** has unfringed petals.

1. *Citrus limon*
2. *Citrus sinensis*
3. *Citrus sinensis*
4. *Anagallis arvensis*
5. *Anagallis monelli*
6. *Arbutus unedo*

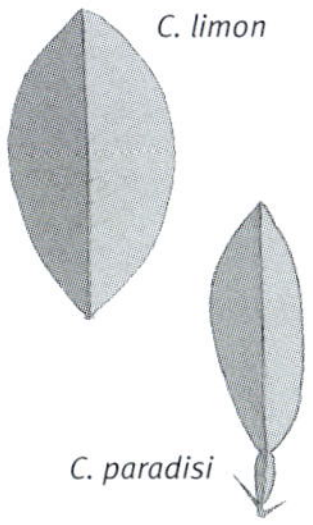

C. limon

C. paradisi

Citrus limon LEMON A familiar evergreen, rounded tree to 4 m, flowering and fruiting throughout the year. Leaves elliptical to lanceolate and shallowly toothed. Flowers male or cosexual, with more than 4 times as many stamens as petals; petals white, often streaked purple. Fruit a lemon, yellow and warty when ripe. Commonly cultivated, rarely naturalised. ***Citrus sinensis*** grows to 10 m and bears flexible spines. Leaves with a rounded base and narrowly winged petiole. Fruit an orange. Very commonly cultivated throughout the region. ***Citrus parasidi*** is similar to *C. sinensis* but with a distinct, broadly winged petiole; fruit a grapefruit.

Order: ERICALES

PRIMULACEAE

Flowers actinomorphic, often 5-parted, the petals fused but appearing free in some species; stamens opposite the lobes; ovary with a single style.

A. arvensis

Anagallis arvensis SCARLET PIMPERNEL A low, weedy annual with prostrate or ascending 4-angled stems. Leaves ovate to lanceolate, opposite and gland-dotted; unstalked. Flowers blue or red, to 7 mm, long-stalked and becoming curved in fruit; petals with a hairy margin, sometimes toothed at the tip. Very common in towns, waste places and near the coast.

A. monelli

Anagallis monelli SHRUBBY PIMPERNEL A low to short perennial with erect or spreading stems, sometimes shrubby and mat-forming in exposed areas. Leaves opposite or whorled, linear-lanceolate to elliptical, unstalked, often with small lateral shoots. Flowers blue (sometimes red, pink or white), to 12 mm across. Stamens with a tuft of hairs at the base. Locally common on stabilised dunes and other coastal sands.

S. valerandi

Samolus valerandi BROOKWEED A hairless perennial herb with a rosette of leaves, and erect flowering stems. Leaves rather shiny, spoon-shaped and scarcely stalked below; stalkless above. Flowers small and white, cup-shaped, to just 3 mm across and with 5 petals. Damp open habitats, often coastal; local.

ERICACEAE

Recent phylogenetic research has resulted in the inclusion of the formerly recognised family Empetraceae (traditionally including genera such as *Corema*) within the Ericaceae. Evergreen trees and shrubs (often characteristically heather-like) with alternate, opposite or whorled leaves. Flowers borne in racemes, panicles or clusters (rarely solitary); sepals 4–5, petals 4–5 and fused, stamens twice as many as petals lobes. Fruit a capsule or berry.

A. unedo

Arbutus unedo STRAWBERRY TREE An evergreen shrub or small tree to less than 12 m. Bark dull brown and fissured. Leaves oblong-lanceolate, short-stalked and somewhat toothed, more or less hairless. Flowers borne in drooping panicles, scented, white, tinged green or pink, bell-shaped with recurved petal lobes, to 9 mm. Fruit a globose berry, ripening deep crimson, rather like a wild strawberry. Common on wooded hill slopes in the Monchique area.

Erica **HEATHERS** Dwarf to medium shrubs with whorled, narrow to linear leaves. Flowers bell-shaped to globular, borne in spikes or panicles, generally 4-lobed. Fruit a dry capsule.

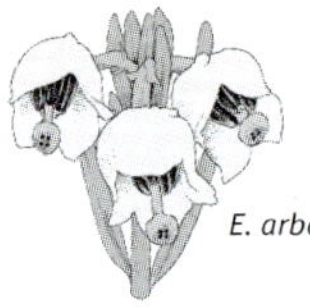
E. arborea

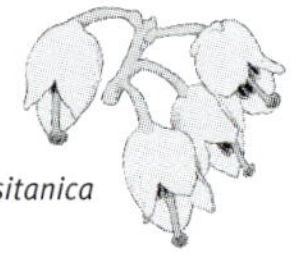
E. lusitanica

Erica arborea **TREE HEATH** A tall shrub or small tree to 4 m with densely hairy younger twigs. Leaves in fours, to 5 mm long. **Flowers pure white,** broadly bell-shaped, to 4 mm long with erect lobes, borne in dense terminal panicles. Common on shaded *matos*, streamsides and on wooded hill slopes. ***Erica lusitanica*** is similar but with larger, paler green leaves (to 7 mm long) and larger flowers (to 5 mm), **white tinged with pink**, particularly in bud, with red stigmas. Habitats similar to those of *E. arborea*, especially in damp places.

E. australis

Erica australis **SPANISH HEATH** A slender shrub to 2 m with suberect branches. Leaves linear, to 6 mm, in whorls of 4, with down-turned margins that completely conceal the lower surface. Flowers 6–9 mm, deep reddish-pink, tubular-bell-shaped with reflexed petal lobes, borne in clusters of 4–8 forming lax, terminal panicles. Common on scrub, fired-damaged *matos* and open woods.

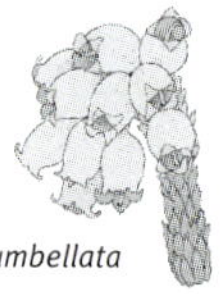
E. umbellata

Erica umbellata Rather similar to *E. australis* but **shorter**, to 80 cm with **smaller flowers** to 4 mm long, borne in terminal umbels of 3–6; corolla rather globose with erect lobes. Heaths, scrub and open woods.

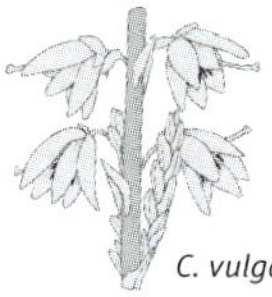
C. vulgaris

Calluna vulgaris **HEATHER** A short, compact scrub to 80 cm; superficially conifer-like when in leaf, typically heather-like when in flower. Leaves scale-like to 3 mm, closely appressed to the stem. Flowers abundant, pink, borne on narrow, crowded racemes; corolla lobed nearly to the base. Frequent in sand-dune plant communities, heaths and moors.

Rhododendron Evergreen shrubs with 5-parted, showy flowers. Commonly planted and naturalised.

Rhododendron ponticum An erect, evergreen shrub with spreading branches. Leaves large, to 20 cm long, entire, leathery, dark green above and paler below; hairless. Flowers showy, borne in lax racemes of up to 15 flowers; corolla pale purple, with 10 stamens. Local on wooded slopes in the Monchique area.

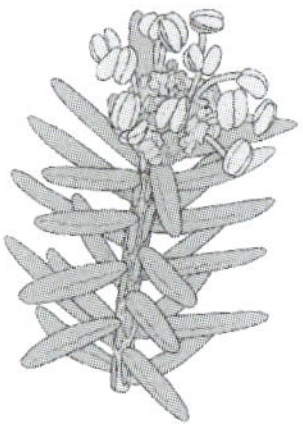
C. album

Corema album A heather-like, erect, dull, brownish-green shrub to 1 m. Leaves to 10 mm, linear with a deep groove running along the centre. Flowers borne in clusters of 5–9, virtually stalkless; male flowers pinkish with red anthers; female flowers with reduced petals or none, and a reddish 3-lobed stigma. Fruit white (rarely pink) and berry-like. Locally common on sand dunes; abundant in some places on Cape St. Vincent.

1. *Erica arborea*
2. *Erica australis*
3. *Calluna vulgaris*
4. *Rhododendron ponticum* PHOTO: TIM RICH
5. *Corema album*
6. *Corema album*

1
2
3
4
5
6

Order: GENTIANALES

RUBIACEAE

A very large family of herbs with opposite or conspicuously whorled leaves; stipules present between each pair of leaves, often leaf-like. Flowers funnel-shaped with a tube, 4–5-parted, borne in dense heads, branched cymes or panicles; ovary inferior. Fruit fleshy or dry and berry-like, 1–2-seeded.

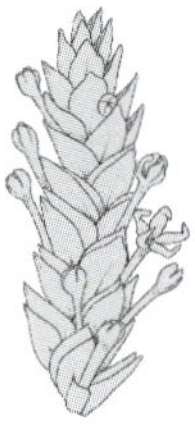

C. maritima

Crucianella maritima CRUCIANELLA A prostrate to spreading low shrub with whitish, hairless stems. Leaves in fours, forming regular, symmetrical, close-set whorls; grey, leathery, spine-tipped with a whitish margin. Flowers yellow, to 13 mm long, 5-lobed. Very common on coastal sands and dunes throughout the region. ***Crucianella latifolia*** is similar but an **annual** with leaves in whorls of 6–8, linear-lanceolate and broader below, not spine-tipped. Flowers cream or purplish, to 7 mm long. Similar habitats to the previous species but rare in the region.

1. *Crucianella maritima*
2. *Rubia peregrina*
3. *Sherardia arvensis*
4. *Valantia muralis*

Asperula Annual or perennial herbs or subshrubs with square stems and leaves in whorls of 4 or more; flowers 4-parted.

A. aristata

Asperula aristata A short to medium subshrub, woody at the base with green or greyish, slightly hairy, 4-angled shoots. **Leaves in whorls only of 4**, lanceolate to linear and colourless at the tip, the margins often down-turned. Flowers pale pink (sometimes greenish or yellowish), funnel-shaped and with 4 petal lobes. Common on the *matos*.

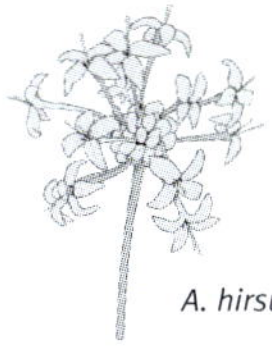
A. hirsuta

Asperula hirsuta A short to medium, scrambling or climbing subshrub, woody at the base with 4-angled, erect stems to 60 cm. **Leaves in whorls of 4 or more**, linear and thin to 1.2 mm wide, **dark green, even when dry**. Flowers pale pink to brownish-red. Common on the *matos*, often supported by other vegetation.

Galium Similar to *Asperula* but with rounded stems, and leaves in whorls of 4 or more. Flowers 4-parted. Fruits with hooked bristles.

G. murale

Galium murale A low to short, typically goose-grass like, sprawling annual with stems to 20 cm, much-branched from the base and sparsely hairy. Leaves narrowly elliptical with a **short spine tip**, in whorls of 4–6. Flowers yellowish, minute to 0.7 mm, borne in lax, few-flowered inflorescences; corolla lobes pointed, erect. Fruit with spreading lobes and hooked bristles. Common on waste and fallow land.

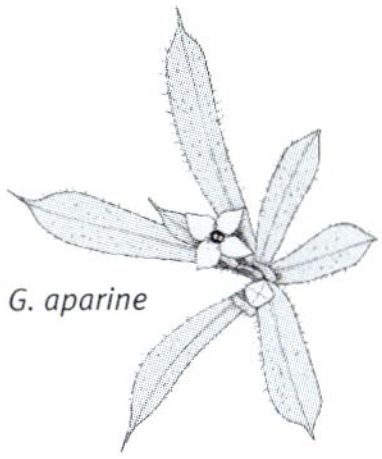
G. aparine

Galium aparine CLEAVERS A low to short, goose-grass like annual with spreading stems to 1 m, often stouter and more hairy at the nodes. Leaves narrowly oblong, in whorls of 6–9. **Corolla to 1.5 mm across, whitish**. Fruit to 5 mm with hooked bristles. A common weed on disturbed ground.

R. peregrina

Rubia peregrina WILD MADDER A medium to tall, trailing or scrambling, hairless, evergreen perennial with a creeping rootstock. Stems square and rough with down-turned bristles. Leaves in whorls of 4–8; oval to elliptical, tough, leathery and dark shiny green, 1-veined. Flowers pale yellow-green, to 5 mm, 5-lobed, forming dense, leafy panicles. Fruit black and fleshy when ripe. Common on low, rocky *matos*.

Sherardia arvensis A small, slender annual with spreading stems to 40 cm or less. Leaves in whorls of 4–6, soon withering below, pale green. Heads 4–10-flowered; corolla pale lilac. Very common on fallow land and in grassy places throughout the region.

S. arvensis

Valantia muralis A low, inconspicuous annual with fleshy, softly hairy stems. Leaves in whorls of 4, ovate and broadest above the middle, to 10 mm long. Flowers greenish-yellow, to 2 mm across, borne in clusters at the base of the leaves. Fruits borne on stalks with conspicuous horn-like appendages. On rocks and old walls.

GENTIANACEAE

Hairless herbs with opposite, untoothed leaves. Flowers 4–5-parted; calyx deeply lobed; petals united in a tube. Fruit a 2-parted capsule. The family is poorly represented in the region.

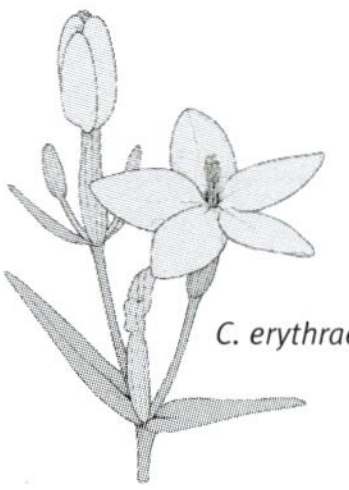
C. erythraea

C. pulchellum

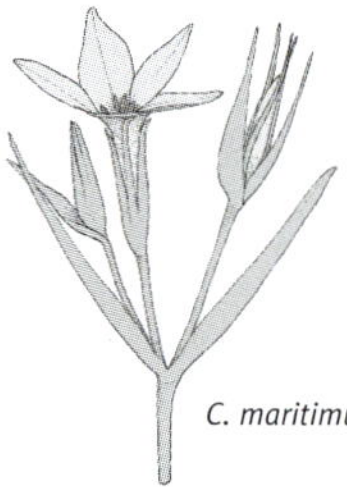
C. maritimum

Centaurium erythraea COMMON CENTAURY A short biennial with a solitary stem that is branched above. Leaves elliptical to ovate, 3–7-veined, the **lower leaves in a distinct rosette**, the upper leaves smaller. Flowers pink, purplish or sometimes white, to 15 mm, virtually stalkless, borne in **flat-topped clusters**. Common in dry grassy places and maritime environments. ***Centaurium pulchellum*** is similar but smaller, lacking a distinct basal leaf rosette, with the stem usually forked in the lower part and with spreading branches, and with fewer flowers borne in lax clusters. Damp, grassy places. ***Centaurium tenuiflorum*** is similar to *C. pulchellum* but the **stem is branched in the upper part**, and the flowers are borne in dense **narrow clusters**. Damp, marshy, coastal places.

Centaurium maritimum YELLOW CENTAURY A short annual to biennial with a solitary stem that is branched above. Leaves fleshy, elliptical to ovate, the lower 2 small, and **not forming a distinct rosette**, the upper leaves much longer. **Flowers pale yellow**, to 18 mm, borne in loosely branched clusters, sometimes solitary; petals lobes 5, elliptical. Sandy and grassy maritime habitats.

Blackstonia perfoliata YELLOW-WORT An erect, hairless, bluish plant to 40 cm tall with stems branching from the base or from the middle, from a basal rosette of broadly ovate leaves that soon wither; stem leaves opposite, ovate-triangular, pointed and narrowed at the base and joined there, **almost encircling the stem**. Flowers yellow, with a short corolla tube and 6–8 spreading lobes; calyx divided into 12 segments. Local in damp habitats or grassy habitats and *matos* following wet winters.

APOCYNACEAE

Trees, shrubs and climbers with opposite, untoothed leaves, often with a milky latex when cut. Flowers 5-parted, the corolla with a distinct tube and 5 lobes. Fruit a pair of follicles, seeds often with hairy tufts.

Nerium oleander OLEANDER A robust, evergreen, upright shrub to 4 m with long, erect branches. Leaves opposite, leathery, linear-lanceolate. Flowers pink, red or white, sweetly scented, borne in showy terminal clusters. Follicles large to 16 cm long. Commonly planted near swimming pools, roadsides and in gardens or as a hedge; naturalised in damp places such as seasonally drying river banks and gullies. Extremely poisonous.

Vinca Perennials with trailing stems and pairs of leathery leaves. Flowers typically periwinkle-like; solitary in the upper leaf axils; corolla propeller-shaped.

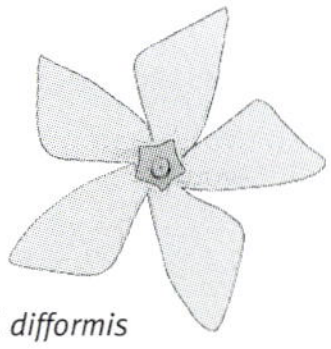
V. difformis

Vinca difformis INTERMEDIATE PERIWINKLE Superficially similar to *V. major* but larger, with stems reaching 2 m and far-spreading, and leaves without a hairy margin. **Flowers pale blue or whitish**, to 70 mm across, with very narrow, hairless or minutely hairy calyx lobes. Common as a wild plant in ditches, roadsides, embankments and shady *matos*.

V. major

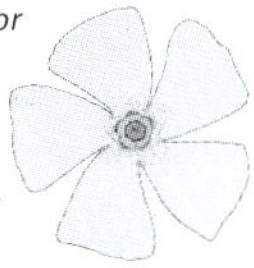

V. minor

Vinca major **GREATER PERIWINKLE** A short to medium, spreading evergreen shrub with long trailing stems, often rooting down. **Leaves large**, to 20 cm (often less), shiny dark green, ovate with a **minutely hairy margin**. Flowers bluish-violet, to 50 mm, the calyx lobes hairy on the margin. Follicles to 50 mm; often absent. Cultivated and naturalised. ***Vinca minor*** is similar to *V. major* but smaller in all parts, with **leaves hairless along the margin**, and flowers to 30 mm across with triangular, hairless calyx lobes. Cultivated and naturalised; common.

1. *Centaurium erythraea*
2. *Blackstonia perfoliata*
3. *Vinca difformis*
4. *Vinca major*

1

2

3

4

The following species were formerly classified under the family Asclepiadaceae, but have since been re-classified as the subfamily Asclepiadoideae in the Apocynaceae:

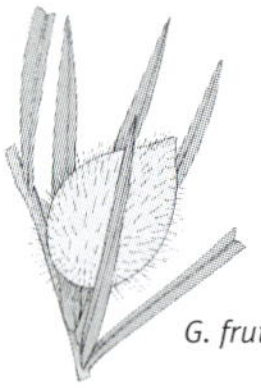

G. fruticosus

Gomphocarpus fruticosus BRISTLE-FRUITED SILKWEED An erect shrub to 1 m with linear-lanceolate, hairless leaves. Flowers white, to 14 mm, borne in hairy, stalked umbels. Fruit very large and conspicuous; an inflated, ball-like, bristly pod to 60 mm long, with cottony seeds. Native to South Africa but planted as an ornamental in developed areas, and possibly naturalised.

O. variegata

Orbea variegata STARFISH FLOWER A short, mat-forming **succulent** with erect and spreading stems to 15 cm high only, without conspicuous leaves. Flowers very distinctive; **star-shaped**, warty, pale yellow with purple spots, foul-smelling to attract flies, borne at the base of the stems, at the perimeter of the plant. Native to South Africa; planted in gardens as an ornamental.

BORAGINACEAE

Family not yet placed in an order and sitting between *Gentianales* and *Solanales*. Herbs or shrubs with often markedly bristly stems and leaves; bristles often with swollen bases. Leaves simple and alternate. Flowers in spiralled clusters, short-stalked, 5-parted, blue in many species. Fruit comprising 4 nutlets, often concealed in a persistent calyx.

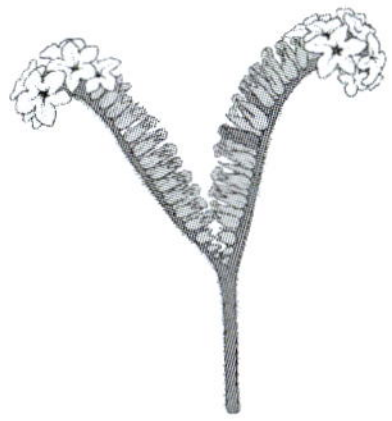

H. europaeum

Heliotropum europaeum HELIOTROPE A low to short, greyish, erect or spreading, much-branched annual. Leaves ovate to elliptical, grey and stalked. Flowers white with a yellow throat, small to 2 mm, borne in distinctly 1-sided, spiralled spikes. Fruit splitting into 4 nutlets. Bare, cultivated and waste ground.

Neatostema apulum YELLOW GROMWELL A short annual with erect, bristly stems that are branched above. Leaves narrowly oblong, and bristly along the margins, **leaves on the stems erect**. Flowers yellow, to 7 mm long. Nutlets pale brown and warted. Common on dry roadsides and *matos*.

L. diffusa

Lithodora diffusa SCRAMBLING GROMWELL A short to medium, bristly, spreading, shrubby perennial to 30 cm. Leaves elliptical, scarcely stalked. Flowers bright blue and funnel-shaped, to 20 mm long, the corolla with a hairy ring in the throat. Nutlets greyish-brown and smooth. Common on coastal sands and cliff tops.

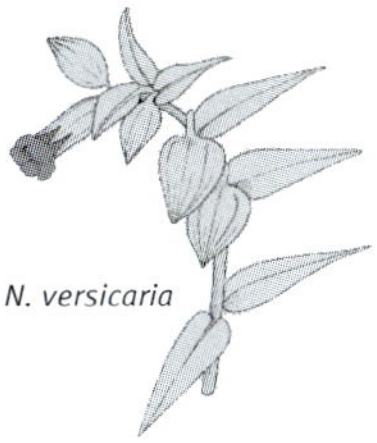

N. versicaria

Cerinthe major HONEYWORT A rather fleshy, grey-green almost hairless annual. Lower leaves oblong, clasping the stem with a heart-shaped base; leaves covered with conspicuous **white swellings**. Bracts ovate to heart-shaped. Flowers dark purple throughout, to 30 mm long, the short petal lobes recurved; borne in a **drooping inflorescence**. Common on coastal sands throughout the region. ***Cerinthe gymnandra*** is sometimes treated as a subspecies but differs in its flowers, which are **white** with a yellow or purple-red base. Habitats and distribution similar to those of *C. major.*

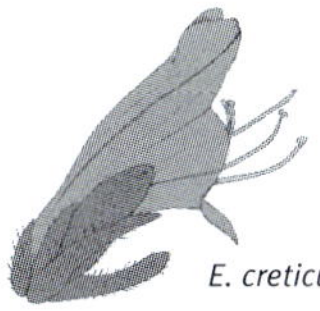

E. creticum

Nonea versicaria A short to medium, greyish, bristly annual with lanceolate leaves, clasping the stem above. Flowers claret-coloured, to 14 mm long, with spreading petal lobes, borne in leafy, 1-sided clusters with conspicuous and enlarged, reddish sepals appearing inflated when in fruit. Nutlets ovoid, ribbed, and constricted above the basal ring. Local in dry, grassy places.

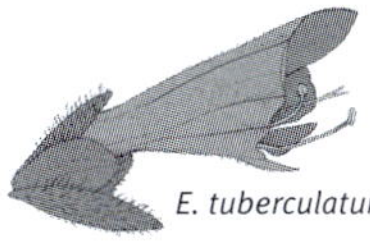

E. tuberculatum

Echium creticum An erect, bristly biennial to 1 m with few stems. Basal leaves narrowly oblong, to 18 cm, with sparse bristles with swollen bases. Stem leaves narrowly elliptical. Flowers pinkish-red, often becoming purple with age, to 40 mm long, broadly funnel-shaped, with 1–2 stamens protruding. Common on roadsides and disturbed grassy places. ***Echium tuberculatum*** is similar to *E. creticum* but with narrowly funnel-shaped **flowers that are dark-purple blue**, and with **2–4 stamens protruding**. Habitats similar to those of *E. creticum*.

1. *Neatostema apulum*
2. *Lithodora diffusa*
3. *Cerinthe major*
4. *Cerinthe gymnandra*
5. *Nonea versicaria*
6. *Echium creticum*

E. vulgare

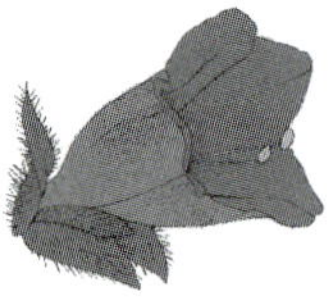

E. plantagineum

Echium vulgare **VIPER'S BUGLOSS** A bristly hairy biennial with one to several stems. Leaves covered in rather soft hairs, elliptical and stalked at the base, narrowly lanceolate and stalkless along the stem. Inflorescence spike-like; flowers to 20 mm, **blue** to blue-purple with **4–5 long-protruding stamens**. Dry open habitats and fallow land. ***Echium plantagineum*** is similar to *E. vulgare* but **softly hairy** (rather than coarsely bristly) with reddish bristles. Leaves with prominent midribs and lateral veins. Flowers to 25 mm, dark blue-purple ageing red, with 2–4 protruding reddish stamens, borne in branched, panicle-like inflorescences. Very common in sandy maritime habitats. ***Echium gaditanum*** is distinguished by its lax inflorescence of clear blue (ageing pink) flowers, in which the corolla tube is abruptly inflated above the middle, with 3–4 stamens protruding. Coastal sands.

Echium candicans **PRIDE OF MADEIRA** A bristly hairy, **large, shrubby perennial** to 2 m tall, with many ascending flowering stems. Leaves dark green, rough, elliptical in shape and pointed at the tips. Flowers bright blue, borne in dense inflorescences on stems leafy below. Native to Madeira but commonly planted in towns, gardens and resorts in the Algarve.

Borago officinalis **BORAGE** A medium, bristly annual with robust, generally branched stems. Basal leaves ovate and light green, stalked; stem leaves smaller and unstalked. Flowers bright blue and star-like with a white centre and a cone of contrasting blackish anthers, to 25 mm, drooping; flowers borne in loosely branched cymes. Very common on cultivated, sandy and waste ground throughout the region.

Anchusa Annual or perennial herbs with blue, purple, white or yellow flowers borne in cymes with bracts. Corolla with 5 hairy scales in the throat, stamens scarcely or not protruding. Nutlets egg-shaped with a thickened basal collar, rough, netted or warted; conspicuous in fruit.

A. calcarea

A. undulata

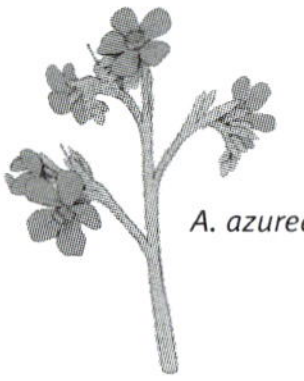

A. azurea

Anchusa calcarea A short to medium, hairy perennial with elliptical leaves forming a bushy basal rosette; upper leaves unstalked and clasping the stem; leaves and stems covered in **bristles with white, pimple-like bases**. Flowers dark blue, funnel-shaped, bracts ovate to lanceolate and **shorter than, or scarcely exceeding the calyx**. Very common on coastal sands, and sandy roadsides. ***Anchusa undulata*** is very similar but with hairs that lack white, pimple-like bases. Dry, sandy places; less common than *A. calcarea* in the Algarve.

Anchusa azurea **LARGE BLUE ALKANET** A robust species to 1 m or more, with dense bristles, often with white, pimple-like bases. Leaves to 30 cm long and 10 cm wide; lanceolate. Flowers large, to 20 mm across, deep blue or purple with a whitish centre (sometimes all white or cream); flowers borne in a large, loose and much-branched inflorescence. Frequent on bare ground.

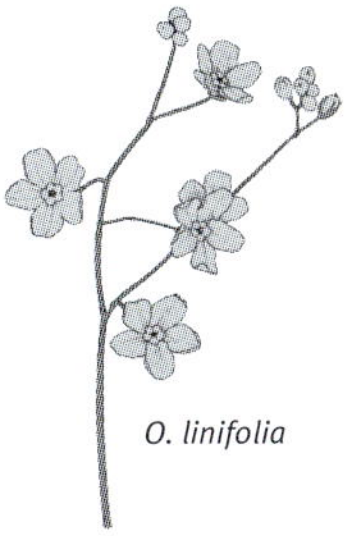

O. linifolia

Omphalodes linifolia ANNUAL OMPHALODES A low to short, erect annual with simple or sparsely branched stems, to 10 cm (sometimes taller). Leaves spatula-shaped and stalked, linear and unstalked above, all with sparsely hairy margins. Flowers white or blue-tinted, borne in long racemes; bracts absent. Common on dry slopes on bare *matos*; often in dense patches.

1

2

3

4

5

7

6

1. *Echium candicans*
2. *Echium plantagineum*
3. *Borago officinalis*
4. *Anchusa calcarea*
5. *Anchusa calcarea*
6. *Omphalodes linifolia*
7. *Anchusa azurea*

Cynoglossum Biennial to perennial herbs with flowers in branched cymes, without bracts. Corolla with a short tube and 5 spreading petal lobes with scales closing the throat; stamens not protruding. Nutlets egg-shaped and with barbed spines.

C. cherifolium

Cynoglossum creticum **BLUE HOUND'S TONGUE** A short, robust, softly hairy biennial with erect, angular stems branched above. Leaves lanceolate, untoothed and **densely hairy to felted on both surfaces**, sometimes clasping the stem. Flowers purplish in bud, opening blue with fine, **darker blue venation**; to 9 mm across, borne in branched cymes that elongate in fruit. Nutlets without a distinct border, and with dense hooked spines. A range of habitats, often on disturbed ground. ***Cynoglossum cherifolium*** is similar but with leaves conspicuously **white-felted** on both surfaces; somewhat undulate. **Flowers reddish-purple** or bluish, borne in crowded racemes that elongate in fruit. Nutlets with a distinctly **thickened border**, and with dense hooked spines, or smooth. Dry, stony places and waste ground. ***Cynoglossum clandestinum*** is similar to both species but with smaller flowers to 6 mm across that are **pale violet** and with **woolly petal lobes**, and nutlets without a distinct border.

1. *Cynoglossum creticum*
2. *Cynoglossum creticum*
3. *Cynoglossum clandestinum*
4. *Cynoglossum clandestinum*
5. *Hyoscyamus albus*

M. laxa

Myosotis laxa **TUFTED FORGET-ME-NOT** A bushy annual to 40 cm. Plant not conspicuously hairy, but with appressed hairs on the stems and leaves. Flowers small, to 4 mm across, pale blue with rounded petal lobes; flowers borne laxly along the stem, densely at the apex. Fruit stalks 2–3 times the length of the calyx. Local in wet habitats.

Order: **SOLANALES**

SOLANACEAE

Herbs or shrubs with simple or pinnate, alternate leaves. Flowers actinomorphic or not, generally 5-parted with a star- or bell-shaped corolla, the petals fused below; stamens 5, attached to the corolla tube; ovary superior with 2 compartments. Fruit a berry or capsule. A well-represented family in the region.

L. europaeum

Lycium europaeum **TEA TREE** A deciduous, **conspicuously and robustly spiny shrub**, to 4 m. Leaves elliptical, broadest above the middle, 20–50 mm long. Flowers pink or white, narrowly funnel-shaped, to 13 mm long, solitary or in small clusters with stamens usually protruding. Fruit a reddish berry. Hedges and thickets. ***Lycium intricatum*** is similar to *L. europaeum* but fleshy with **smaller leaves** (5–15 mm long), and slightly larger flowers (to 18 mm long), lilac, pink or white with **stamens not protruding**. Fruit an orange-red or blackish berry. Local in sandy habitats near the coast.

L. barbarum

Lycium barbarum **DUKE OF ARGYLL'S TEAPLANT** A deciduous shrub to 2.5 m with arched, grey-white branches with few, slender spines. Leaves alternate or in clusters, narrowly elliptical and broadest at the middle; untoothed, to 10 cm long. Flowers red-purple, turning brown with age, trumpet-shaped, **small** to 9 mm long; or in small clusters of 1–3 with stamens long-protruding. Fruit a orange-red oval berry. Occasionally planted as a hedge plant in coastal areas in the region.

H. albus

Hyoscyamus albus **WHITE HENBANE** A sticky annual or short-lived perennial with long, sparsely branched stems to 90 cm, often woody below. Leaves to 10 cm long, broadly oblong and wedge-shaped or heart-shaped at the base, with wide teeth along the margin. **Flowers stalkless**, at least above, borne in long, dense, one-sided spikes; calyx densely hairy and swollen below, ending in short, triangular teeth; corolla to 30 mm, greenish or **yellowish-white** with a green or purple throat; **stamens not, or scarcely protruding**. Local on waste ground and near buildings.

Hyoscyamus niger **HENBANE** A sticky, unpleasant-smelling annual or biennial with long, laxly branched stems to 90 cm, often woody below. Leaves to 10 cm long, ovate and coarsely toothed, forming a basal rosette; stalked below but stalkless above and somewhat **clasping the stem**. Flowers virtually stalkless, borne in long, dense, one-sided spikes; calyx densely hairy and swollen below (especially in fruit), ending in short, triangular teeth; corolla to 30 mm, slightly uneven (zygomorphic), **dull pale yellow with purple veins**; stamens slightly protruding. Fruit a capsule opening at the top. Disturbed waste places; a casual weed in the region.

P. ixocarpa

Physalis ixocarpa A more or less hairless, branched, erect annual to 60 cm. Leaves lanceolate, with course teeth or entire; stalked. Flowers solitary in the leaf axils; calyx ovate, green and purple-veined, with 5 teeth; corolla to 30 mm in diameter, bright yellow with brownish markings in the centre; filaments purple. Fruit borne in a 'Chinese lantern' (a much inflated calyx enclosing a berry). Native of North America; naturalised in the region.

Solanum **NIGHTSHADES** Herbs or shrubs with flowers in 1-sided cymes or umbels; leaf-opposed. Corolla star-shaped with spreading petal lobes; stamens protruding, forming a cone around the stigma. Fruit a berry. Usually poisonous.

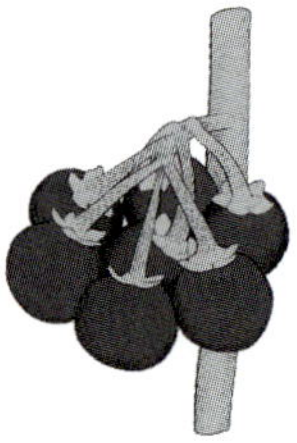

S. nigrum

Solanum nigrum **BLACK NIGHTSHADE** A low to medium, hairless to hairy **annual**; stems spreading and blackish. Leaves ovate to lanceolate, toothed, lobed or entire; stalked. Flowers white and star-like with a yellow cone of anthers, to 14 mm borne in clusters of 5–10. Berry round and green ripening matt-black; to 10 mm. A common weed on cultivated land; frequent in orange groves and farms in the region.

Solanum linnaeanum **APPLE OF SODOM** A small to large, very **spiny shrub** to 3 m with much-branched, stout stems covered in yellow spines; slightly hairy. Leaves ovate and pinnately, bluntly lobed; prickly, and stalked. Flowers violet, to 30 mm across, pentagonal in shape, solitary or in clusters. **Berry large** and spherical, to 30 mm in diameter, marbled green and white, later yellow or brown and shiny; rather tomato-like. Local on coastal roadsides and waste places and near derelict buildings in the region.

S. lycopersicum

Solanum lycopersicum **TOMATO** A glandular-hairy, strongly scented annual. Leaves to 20 cm with ovate leaflets. Cymes with up to 20 yellow flowers. Fruit an edible red or yellow berry (tomato). Cultivated and naturalised as a casual weed.

S. tuberosum

Solanum tuberosum **POTATO** A sparsely hairy, bushy annual with underground stolons bearing terminal tubers (potatoes). Leaves divided with up to 7 pairs of ovate leaflets. Cymes many-flowered with pentagonal purple or white flowers with yellow cones of anthers. Cultivated but probably not naturalised.

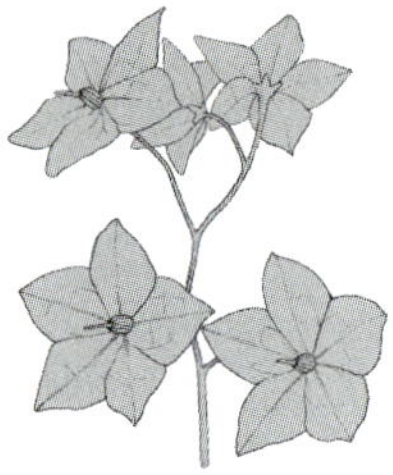

S. jasminoides

Solanum jasminoides **POTATO VINE** An ornamental evergreen **climber**. Leaves bright green, leathery and shiny, pointed and lanceolate with heart-shaped base; stalked. Flowers white and star-like with yellow cones of anthers, borne in lax, many-flowered, showy cymes. Widely planted in developed areas; not naturalised.

Mandragora autumnalis **MANDRAKE** A ground-hugging, stemless, hairless or hairy perennial with a robust rootstock. Leaves forming a large, flat **rosette** to 50 cm across; ovate, bright green and shiny, stalked and distinctly wavy-margined. Flowers purple, to 40 mm across with 5 triangular lobes, borne in the centre of the rosette. Fruit an orange or yellow, egg-shaped berry to 30 mm long, held in a large, persistent calyx that equals or exceeds the fruit. Fallow land and bare places. The roots can be extremely toxic and were formerly used in the Mediterranean to relieve pain and to induce sleep.

Datura Erect annuals with actinomorphic, showy, trumpet-shaped flowers. Fruit a large, spiny capsule.

D. stramonium

D. inoxia

Datura stramonium A tall and vigorous, unpleasant-smelling, normally **hairless** annual to 1.5 m with stout, spreading stems. Leaves ovate to elliptical, lobed with jagged teeth. Flowers white, sometimes flushed with purple and trumpet-like, to 10 cm long, solitary in the leaf axils of the upper leaves; calyx large, to half the length of the corolla, sharply angled. Fruit an erect, large, spiny capsule to 15 mm long. Sporadic on bare waste ground, coastal roadsides, and cultivated land. ***Datura inoxia*** is similar but softly **hairy**, usually smaller, to 50 cm, with **larger flowers** to 19 cm (but usually smaller), and **capsules nodding** when mature. Habitats similar to those of *D. stramonium*. ***Datura metel*** may occur in the region, but is regularly confused with *D. inoxia* to which it is similar, but **hairless** or nearly so. Double-flowered forms are cultivated.

1. *Solanum nigrum*
2. *Solanum lycopersicum*
3. *Solanum tuberosum*
4. *Solanum linnaeanum*
5. *Datura inoxia*

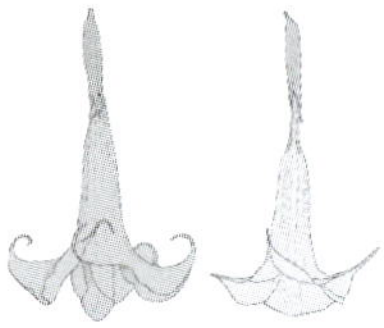

B. versicolor *B. arborea*

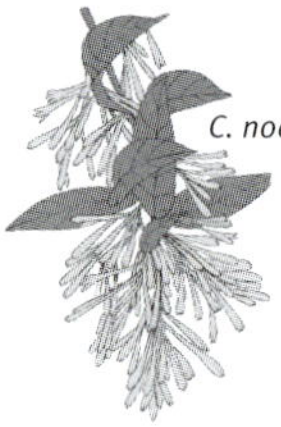

C. nocturnum

Brugmansia versicolor **ANGEL'S TRUMPET** An ornamental shrub to 5 m with thick, stiff stems. Leaves ovate, stalked and untoothed. Flowers very showy; white flushed with pale orange or pink, trumpet-like, **very large** to 20 cm long and fragrant; solitary in the upper leaf axils; calyx inflated. Fruit large and egg-shaped. Endemic to Ecuador; widely cultivated. ***Brugmansia arborea*** is similar but a larger **tree** with **white**, pale greenish or cream flowers.

Cestrum nocturnum An unpleasant-smelling shrub with alternate, simple, lanceolate, pointed leaves. Flowers abundant and showy; tubular, white or cream and greenish, with 5 stamens not protruding. Fruit a 2-parted berry. Planted in the region, possibly naturalised.

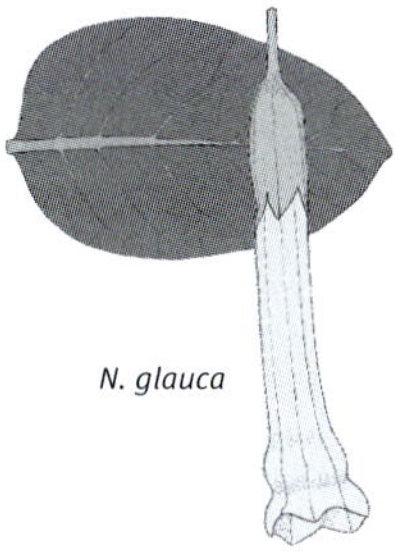

N. glauca

Nicotiana glauca **SHRUB TOBACCO** A hairless, soft-wooded shrub or small tree to 6 m with lax, grey branches. Leaves elliptical to lanceolate, pointed, grey-green and stalked. Flowers greenish-yellow, to 40 mm long, borne in lax panicles; corolla tubular. Fruit an egg-shaped capsule to 10 mm long, exceeding the persistent, papery calyx. Cultivated and naturalised in dry waste places.

Petunia axillaris is a short, bushy, sticky-hairy plant with large, white, trumpet-like flowers that are scented at night. Commonly planted.

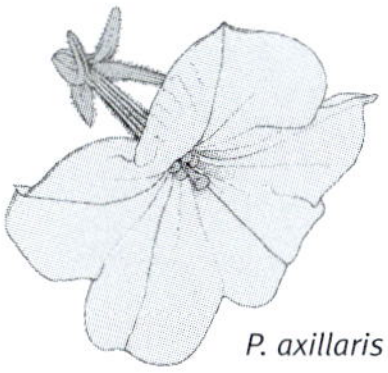

P. axillaris

Solandra maxima **HAWAIIAN LILY** A shrubby climber with ovate, dark green leaves and **large, showy, trumpet-shaped flowers to 20 cm long**; yellow-cream with purple veins and prominent stamens. Fruit a berry. A striking exotic ornamental native to Central and South America, widely planted in towns, gardens and resorts in the Algarve.

Capsicum annuum **CHILLI** A domesticated plant native to the Americas; widely cultivated in the Algarve as a source of culinary chillies.

Salpichroa origanifolia A much-branched, hairy perennial with woody stems below, usually scrambling among surrounding vegetation. Leaves ovate, short-stalked and entire. Flowers white and bell-shaped, to 10 mm long. Fruit a small cream-white berry. Native of South America; naturalised on roadsides and in grassy waste places near Burgau and possibly elsewhere.

CONVOLVULACEAE

Climbing herbs with twining stems. Flowers cosexual and 4–5-parted. Fruit a 2–4-valved capsule.

C. epithymum

Cuscuta epithymum **COMMON DODDER** A spreading, spaghetti-like, parasitic plant without chlorophyll. **Stems reddish** and thread-like, entwined around and attached to the stems of green plants. Leaves reduced to tiny scales. Flowers pale pink with a reddish calyx, to 4 mm, scented and 5-parted with spreading petals borne in ball-like clusters. Parasitic on herbs and shrubs in the Leguminosae and Ericaceae in this region.

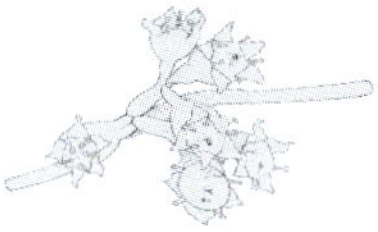

C. campestris

Cuscuta campestris **FIELD DODDER** Similar to *C. epithymum* but with **yellow-orange stems** and smaller, 5-parted flowers, to 2 mm across, borne in lax clusters. Native to North America but naturalised and parasitic on various herbs, particularly on Apiaceae in this region. ***Cuscuta europaea*** is similar to *C. epithymum* but with stouter purplish or reddish stems and **4-parted flowers** borne in larger clusters, to 12 mm across. Parasitic on various hosts from many families. Rare. ***Cuscuta suaceolens*** is distinguished by its grey-purple stems and by its 5-parted, white flowers, which often have greenish sepals, are **stalked**, and are borne in **lax (not tight) clusters**. Parasitic on legumes and other herbs.

---

1. *Brugmansia arborea*
2. *Nicotiana glauca*
3. *Solandra maxima*
4. *Salpichroa origanifolia*
5. *Cuscuta campestris*

Calystegia Perennial herbs with twining or prostrate stems with a white latex. Flowers large and broadly funnel-shaped.

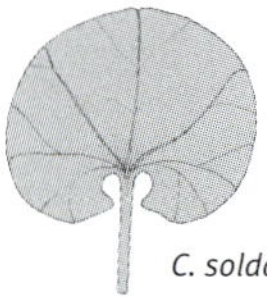

C. soldanella

Calystegia soldanella **SEA BINDWEED** A low, hairless, prostrate, spreading perennial. Leaves kidney-shaped, deep green and veined, rather fleshy and long-stalked. Flowers pale pink with white stripes, to 50 mm; solitary. Very common on coastal dunes.

C. althaeoides

Convolvulus althaeoides **MALLOW-LEAVED BINDWEED** A trailing or twining, hairy perennial with a slender, creeping rootstock. Leaves mallow-like; greyish, heart-shaped below and stalked, deeply lobed above with linear divisions; all hairy with somewhat wavy margins. Flowers pale pink with a deep pink throat, to 50 mm across, solitary or paired, and long-stalked. Very common on roadsides, sandy waste places and *matos*.

C. arvensis

Convolvulus arvensis **BINDWEED** A creeping or twining, more or less hairless perennial, with arrow- to oblong-shaped, stalked leaves. Flowers pale pink to white with paler stripes, to 30 mm across, solitary or paired on stalks shorter than their leaves; scarcely scented. Frequent on fallow land and waste ground.

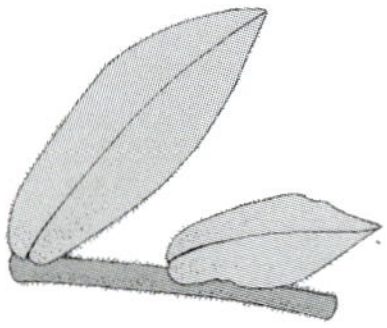

C. tricolor

Convolvulus tricolor **DWARF CONVOLVULUS** A hairy, short, spreading annual. Leaves ovate to elliptical, untoothed and unstalked. **Flowers conspicuously with 3 bands of colour**: yellow in the throat, then white, and finally bright blue on the perimeter; to 50 mm across, solitary or paired on short stalks. Capsule hairy. Frequent on bare and sandy ground and waste places near the sea. ***Convolvulus meonanthus*** is similar to *C. tricolor* but with narrower (almost linear) leaves, smaller flowers only 14–22 mm across, and a hairless capsule. Cultivated ground. ***Convolvulus pentapetaloides*** is similar but smaller still; the flower less than 10 mm across and distinctly 5-lobed. Dry places in the east of the region. ***Convolvulus siculus*** is similar to *C. pentapetaloides* but with leaves lanceolate to ovate, untoothed and **stalked**. Flowers blue with a yellowish centre, 9–12 mm across, distinctly 5-lobed and solitary or paired on short stalks. Dry, bare places.

C. pentapetaloides

Ipomoea acuminata **MORNING GLORY** A perennial herb with long trailing stems. Leaves to 16 cm, ovate and **entire** to deeply 3-lobed, heart-shaped at the base. Flowers large and conspicuous: sky blue or purplish (rarely white) but fading to pink, to 85 mm. Naturalised in thickets, usually near towns. ***Ipomoea hederacea*** is an exotic ornamental native to tropical America with **3-lobed leaves** and light purple flowers. Planted in developed areas.

I. hederacea

Ipomoea batatas **SWEET POTATO** A herbaceous perennial vine with lobed leaves and typically convolvulus-like, purplish, trumpet-shaped flowers. The large reddish tubers are cultivated as a food source. Cultivated occasionally in gardens in the region.

1. *Calystegia soldanella*
2. *Convolvulus althaeoides*
3. *Convolvulus arvensis*
4. *Convolvulus tricolor*
5. *Jasminum fruticans*

Order: LAMIALES

OLEACEAE

Trees and shrubs, generally with opposite, simple leaves. Flowers in cymes or panicles, 4-parted; calyx with 4 small teeth; corolla with 4 free or fused petals; stamens often 2. Fruit a capsule, berry or winged nut.

J. fruticans

Jasminum fruticans **WILD JASMINE** An evergreen shrub to 3 m with slender, 4-angled stems. **Leaves alternate**, usually trifoliate, shiny, with ovate to lanceolate leaflets. Flowers yellow, to 15 mm, unscented, borne in clusters of up to 5. Fruit a black, shiny berry. Common on the *matos* and on waysides.

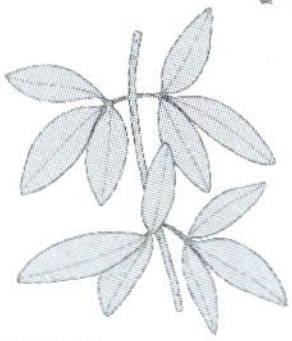

J. mesnyi

Jasminum mesnyi A scrambling, evergreen shrub with angular stems. Leaves opposite, deep green, trifoliate with elliptical leaflets. Flowers borne singly and laterally, bright yellow, often with an orange-tinged centre, to 40 mm, with 6 or more overlapping, oval petals. Widely cultivated.

F. angustifolia

Fraxinus angustifolia **NARROW-LEAVED ASH** A deciduous tree to 25 m with grey, fissured bark. Buds dark brown, twigs hairless and flattened at the nodes. Leaves opposite and pinnate with 5–13 oblong toothed leaflets. Flowers lack both calyx and corolla. Fruit typically ash-like (a samara), hairless, light green and single-winged. River banks; local.

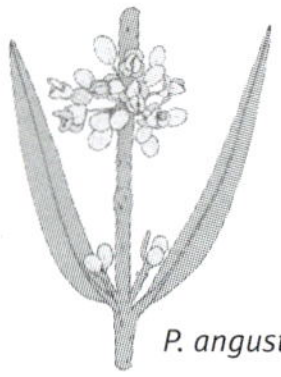

P. angustifolia

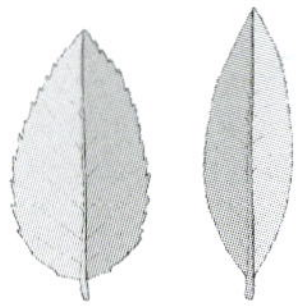

P. latifolia

1. *Olea europaea*
2. *Phillyrea angustifolia*
3. *Scrophularia frutescens*
4. *Scrophularia scorodonia*
5. *Verbascum thapsus*

Olea europaea **OLIVE** A familiar, much-branched tree to 15 m with a grey trunk. Leaves opposite; grey-green, silvery beneath, minutely scaly, lanceolate, untoothed and short-stalked. Flowers small, whitish, borne in erect clusters. Fruit an olive. An important species that is extensively cultivated, and very commonly naturalised on the *matos*. The wild variety, **var. *sylvestris***, differs from the cultivar (**var. *europaea***) in having spiny lower branches.

Phillyrea angustifolia An olive-like evergreen shrub to 2 m with upright branches and grey bark. Leaves opposite, linear to lanceolate, untoothed to finely toothed, with 4–6 pairs of **obscure veins**. Flowers greenish, small (to 3 mm), 4-lobed and borne in lateral clusters to 10 mm across, **fragrant**. Fruit small to 8 mm, fleshy with a point, blue-black when ripe. Very common on the *matos* and fixed dunes throughout the region. ***Phillyrea latifolia*** is similar but with finely hairy twigs and **2 types of leaf** — juvenile leaves ovate and heart-shaped at the base, adult leaves elliptical — both leathery dark green, toothed or not, and with 7–10 pairs of **distinct lateral veins**. Flowers whitish, fruit similar to those of *P. angustifolia*. Less frequent than *P. angustifolia* but in similar habitats.

1

2

3

4

5

Ligustrum lucidum CHINESE PRIVET A small, evergreen tree to 10 m tall with opposite, glossy, dark green leaves. Flowers typically privet-like: white, small and borne in dense panicles to 20 cm long; corolla tubular with 4 spreading lobes. Berry black when ripe, to 12 mm long. Native to China and widely planted.

SCROPHULARIACEAE

This family's circumscription has been dramatically altered in light of phylogenetic analyses, which identified the traditional Scrophulariaceae to be polyphyletic. Many genera have since been transferred to the families Plantaginaceae and Orobanchaceae.

Herbs (rarely shrubs or trees) with opposite or alternate leaves. Flowers zygomorphic, usually in spikes or racemes; bracts present; calyx 4–5-lobed or 2-lipped; corolla 5-lobed and/or markedly 2-lipped; stamens 2 or 4. Fruit usually a 2-parted capsule.

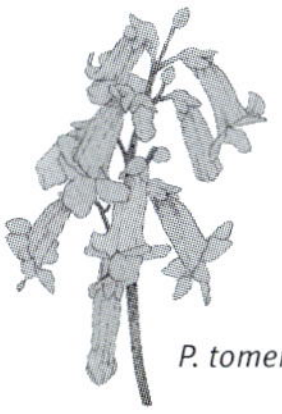

P. tomentosa

Paulownia tomentosa FOXGLOVE TREE A deciduous, rather lax tree with a rounded top to 16 m. **Leaves large** and opposite, ovate, 3–5-lobed, long-stalked, wrinkled and downy. Flowers bluish-purple and funnel-shaped, 2-lipped, to 50 mm long, borne in terminal clusters. Fruit a 2-parted, pointed capsule to 50 mm long. Planted on roadsides in the Monchique area.

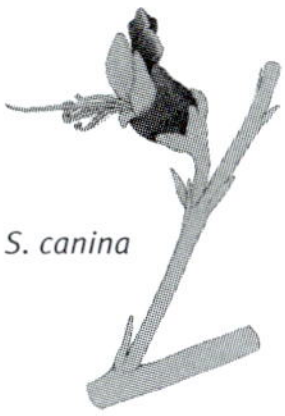

S. canina

Scrophularia canina FRENCH FIGWORT A medium, hairless, much-branched perennial. Leaves below pinnately, narrowly lobed with teeth; upper leaves elliptical and toothed, usually **alternate**. Bracts small and **not leaf-like**. Flowers small, to 5 mm long, purple and white, and borne numerously in lax terminal, cylindrical clusters; calyx lobes with a membranous margin. Dry rocky places.

S. frutescens

Scrophularia frutescens A medium, hairless, sparingly branched perennial, woody below; rather similar to *S. canina* in appearance. **Leaves mostly opposite**, lanceolate, entire or toothed and leathery. **Bracts leaf-like**, at least below. Flowers small, to 4 mm long, dull purple, and borne numerously in lax terminal, cylindrical clusters. Very common on maritime sands in the region.

Scrophularia scorodonia BALM-LEAVED FIGWORT A medium to tall, more or less **downy** perennial. **Leaves entire, with toothed margins**; slightly wrinkled, ovate, pointed. Flowers purple, to 12 mm long, borne in erect, sparse inflorescences; **calyx lobes with broad membranous margins**. Local in woods and in damp habitats on higher ground.

Verbascum Herbs with large basal rosettes, and often grey-hairy leaves. Flowers usually yellow or white with a 5-parted equally lobed calyx.

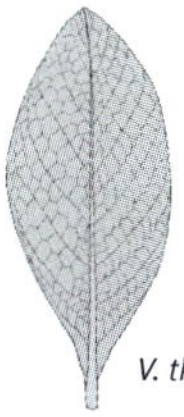

V. thapsus

Verbascum thapsus GREAT MULLEIN A tall, white-woolly biennial to 2 m. Basal leaves elliptical, rather wavy-margined and blunt, toothed or not, with a narrow, winged stalk; **stem leaves with winged bases running down the stem (decurrent) almost to the leaf beneath**. Flowers yellow, to 35 mm across, borne in an often solitary, woolly, **dense terminal spike-like inflorescence**; calyx to 12 mm; petals bright yellow with 5 more or less equal lobes; stamens 5, the upper 3 with hairy filaments, the **lower 2 almost hairless**. Local in a range of dry habitats.

V. barnardesii

V. sinuatum

Verbascum barnadesii A stout, erect, medium, glandular, white-woolly biennial to 1 m. **Basal leaves lobed**, narrowly lanceolate, with toothed margins, short-stalked. Inflorescence solitary, **lax** and spike-like. Flowers bright yellow, to 35 mm across, often with purple-brown markings; **calyx small (to 4 mm)**. Dry, rocky places.

Verbascum sinuatum A stout, erect, medium, **grey or yellow-woolly** biennial to 1 m. Basal leaves large and dense, forming distinctive rosettes, **conspicuously wavy-pinnately lobed**. Flowers yellow, to 30 mm across, borne in clusters on a widely branching inflorescence; stamens with violet-hairy filaments. Rocky, grassy and coastal habitats; common.

Myoporum A genus of trees and shrubs native to Australasia, formally classified in the Myoporaceae. Flowers 5-parted. Fruit a berry.

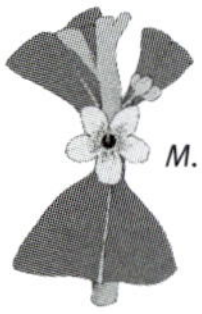

M. laetum

Myoporum laetum An evergreen shrub or tree to 15 m with sticky shoots. Leaves alternate, narrowly lanceolate, slightly toothed towards the tip, hairless and dotted with oil glands. Flowers borne in lateral clusters of 2–6, white with purple markings, to 12 mm; stamens 4. Fruit a purple berry. Sometimes planted the region.

ACANTHACEAE

Herbs with simple, often lobed leaves and erect, (usually) unbranched stems. Bracts conspicuous and spiny. Flowers borne in dense spikes; calyx 4-lobed and 2-lipped, corolla zygomorphic, 1–2-lipped, the lower lip 3-lobed; stamens 4, not protruding. Fruit a capsule.

A. mollis

Acanthus mollis BEAR'S BREECHES A robust, bushy perennial to 1.5 m. Leaves large, shiny dark green, pinnately lobed, soft, hairless or nearly so, and long-stalked. Stem leaves small and few. Bracts purple-tinged and spiny. Flowers white, to 50 mm long, borne in dense spikes; corolla 3-lobed; calyx purple and hairless. Mainly in the hills in damp, shady places in the region, sometimes on roadsides or planted at lower altitudes; fairly common.

VERBENACEAE

Herbs and shrubs with opposite leaves. Flowers in clusters or heads; corolla comprising a slender tube and flat limb, often 2-lipped. Fruit 2- or 4-parted, berry-like or nutlets.

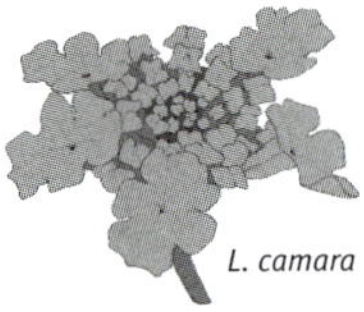

L. camara

Lantana camara LANTANA A small prickly shrub to 1.5 m with square and prickly branches that are hairy when young. Leaves opposite, ovate, toothed and short-stalked. Flowers bright yellow or orange, ageing red, to 5 mm, congested in tight, often paired heads carried on long stalks; corolla slightly 2-lipped. Fruit a small black berry. Cultivated in gardens and possibly naturalised.

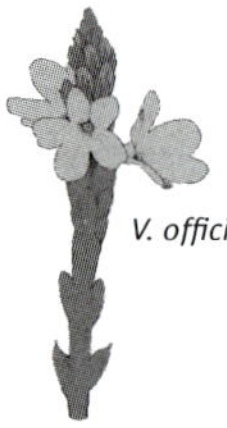

V. officinalis

Verbena officinalis VERVAIN A medium, rough-hairy perennial with slender, stiffly erect and square stems; superficially like a Labiate. Leaves opposite, lanceolate to diamond-shaped in outline but deeply pinnately divided once or twice; stalked below; stalkless and often unlobed above. Flowers pink, to 5 mm, slightly 2-lipped, borne in long, slender, leafless spikes to 30 cm long. Fruit separating into 4 nutlets. Bare waste ground.

BIGNONIACEAE

A large, tropical and subtropical family of trees and climbers with opposite leaves and large, tubular flowers with 4 stamens in 2 pairs. Fruit a pod-like capsule splitting lengthways into 2 halves, containing numerous, often winged seeds.

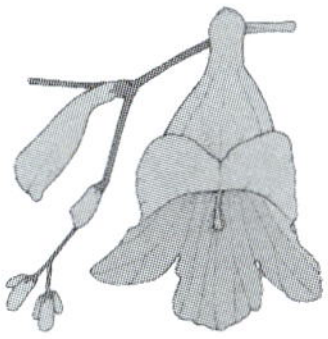

J. mimosifolia

Jacaranda mimosifolia **JACARANDA** A small tree to 8 m. Leaves twice pinnate with numerous, opposite, ovate to diamond-shaped, downy leaflets. Flowers blue, trumpet-shaped, to 60 mm long, 2-lipped; borne in showy, terminal panicles which are borne in summer, before the leaves have developed. Pod ovate, brown and papery when ripe, to 80 mm long. Frequently planted along roadsides in towns.

LAMIACEAE

A large and important family of herbs and shrubs that is well represented in the area; often glandular and aromatic. Leaves opposite and simple. Flowers zygomorphic, often borne in whorls around the stem; calyx 5-lobed, corolla 2-lipped. Fruit comprising 4 nutlets concealed within the persistent calyx.

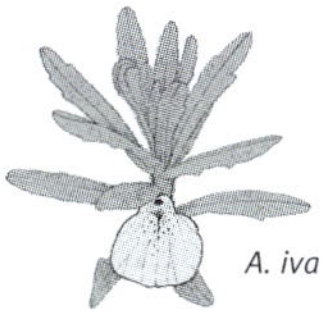

A. iva

Ajuga iva **SOUTHERN BUGLE** A tufted or sprawling, low to short, softly hairy perennial. Leaves broadly linear, normally with short lobes, sometimes unlobed. Bracts similar to the leaves and exceeding the flowers. Flowers white or cream to pale purple, often with small darker spots; to 14 mm long, the **upper lip entire and highly reduced**. Common on low, stony ground.

1. *Verbascum sinuatum*
2. *Acanthus mollis*
3. *Jacaranda mimosifolia*
4. *Verbena officinalis*
5. *Ajuga iva*

Teucrium Flowers with a 2-lipped calyx, and corolla with a single 5-lobed lower lip.

T. fruticans

T. pseudochamaepitys

Teucrium fruticans TREE GERMANDER A spreading, white-felted evergreen shrub to 2 m with square stems. Leaves lanceolate, flat, short-stalked, entire and white-felted; bracts leaf-like. Flowers paired, blue or lilac, to 25 mm with long-protruding and conspicuous stamens; calyx bell-shaped. Coastal *matos* and planted.

Teucrium pseudochamaepitys A sparsely branched, woolly perennial to 40 cm, woody at the base. Lower leaves once or twice divided, the upper 3-parted, all with linear, untoothed lobes. Inflorescence unbranched, the bracts equalling or exceeding the flower stalks. Calyx bell-shaped, glandular-hairy with triangular teeth exceeding their tube. **Corolla twice as long as the calyx**, to 15 mm, whitish and marked, with long-protruding stamens. Common on the *matos*.

1. *Teucrium fruticans*
2. *Teucrium pseudochamaepitys*
3. *Teucrium polium*
4. *Prasium majus*
5. *Marrubium vulgare*

T. spinosum

T. polium

Teucrium spinosum **SPINY GERMANDER** A **glandular-hairy** or woolly annual to 50 cm. Stems much-branched and more or less spiny and leafless when in flower. Lower leaves oblong, narrowed at the base, sharply toothed, the uppermost small and untoothed. Flowers white, to 8 mm long and held **upsidedown** (resupinate) in small whorls; calyx with spiny teeth. Fallow land or sandy places.

Teucrium polium **FELTY GERMANDER** A dense, dwarf shrub to 30 cm tall, densely white-, yellow- or grey-**felted**. Leaves 7–27 mm long, narrowly oblong, often with down-turned margins, bluntly toothed; bracts leaf-like but untoothed. Flowers white or pink-tinted, to 5 mm long, borne in felty, rounded **heads of up to 20 flowers**. Low coastal *matos*. **Subsp. *vicentinum*** has **long leaves** to 28 mm, densely yellow- or grey-felted stems, and hairless corolla lobes. Rocky cliff-top *matos* on Cape St. Vincent. ***Teucrium haenseleri*** is similar but with reddish or green **unfelted** but sparsely hairy stems, narrower leaves (often deflexed), and heads of 3–10 whitish flowers with triangular corolla lobes.

Prasium majus **PRASIUM** A narrow, erect shrub to 1 m, hairless or slightly hairy. **Leaves shiny** dark green, ovate, pointed and toothed, heart-shaped at the base, stalked; bracts similar. Flowers white, to 20 mm long, borne in terminal racemes; calyx 2-lipped with bristle-tipped teeth; corolla with the middle lobe the largest. Nutlets shiny black when ripe. Very common on the *matos*.

Marrubium vulgare **WHITE HOREHOUND** An erect, medium, white-downy, slightly aromatic, rather nettle-like perennial. Stems erect, square and branched. Leaves ovate, heart-shaped at the base and wrinkled on the surface, with stalks shorter than their blades. Flowers small and rather insignificant; the corolla 2-lipped and white, to 15 mm, borne in dense whorls in the leaf axils up the stem; calyx with 10 equally short, hooked teeth. Common on rocky and stony ground, also on the *matos*.

Sideritis Erect, aromatic herbs, perennials or shrubs. Calyx bell-shaped, 5-toothed and 10-veined; corolla usually yellow and 2-lipped without protruding stamens.

S. romana

S. hirsuta

Sideritis romana A shaggily hairy, **short** annual to 30 cm. Leaves oblong, toothed and mostly unstalked; bracts similar. Flowers yellow, sometimes purple or white, to 10 mm long, borne in distant whorls; corolla upper lip flat and undivided; **calyx strongly veined** and with spreading teeth when in fruit. Local in various dry, scrubby habitats.

Sideritis hirsuta A tall, hairy perennial to 50 cm with at a least some spreading hairs on the stems. Leaves ovate, to 20 mm long with a sharply toothed margin; green or greyish. Flowers yellow with a paler or whitish upper lip (not dark-tinged within), to 10 mm long, borne in distant whorls; calyx actinomorphic and **not 2-lipped**. Local in dry, scrubby habitats.

Sideritis linearifolia A more or less hairless, rather tall, erect perennial to 60 cm. Stems greenish, 4-angled, with forwardly curved hairs on 2 opposite sides. Leaves to 30 mm long, narrowly lanceolate and entire with spine-tipped teeth. Lowermost bracts with a long terminal tooth. Flowers borne in congested clusters of about 6; calyx with a ring of hairs inside; corolla to 10 mm, yellow, often purple-tinged. Locally common on cliff tops and on roadsides, especially in the Cape St. Vincent area.

Phlomis Herbs or shrubs with 2-lipped flowers; the upper lip notably hooded and notched at the tip, the lower lip 3-lobed; stamens protruding or not.

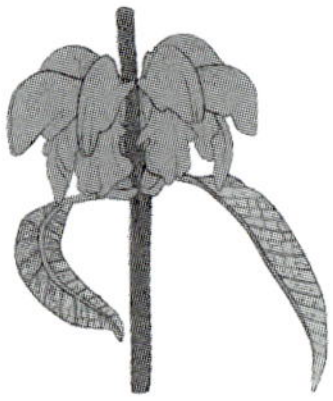

P. purpurea

P. herba-venti

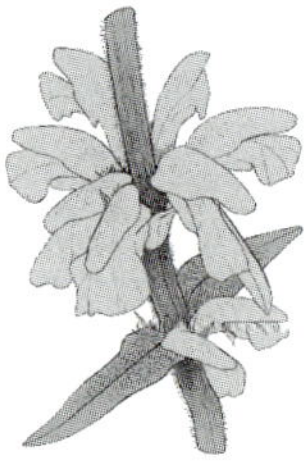

P. lychnitis

Phlomis fruticosa **JERUSALEM SAGE** A large, sage-like, grey-felted, erect or spreading, lax evergreen shrub to 1.5 m. Leaves elliptical and usually untoothed, thick, stalked, greyish above and white-felted below. Flowers large and yellow, to 35 mm, borne in dense whorls. Planted; possibly naturalised.

Phlomis purpurea A sage-like, grey-felted, erect and lax evergreen shrub to 2 m; hairy but not glandular. Leaves thick, leathery and wrinkled; greyish above and white-felted beneath; all leaves stalked. Flowers pale purple (sometimes white), borne in congested clusters of 10–12; to 26 mm long, borne in distant whorls; calyx grey-felted. Very common on the *matos*. ***Phlomis herba-venti*** is similar to *P. purpurea* but **smaller**, to less than 70 cm with **leaves that are scarcely hairy or hairless above**, but white-felted beneath, and calyx lobes with bristle-tipped teeth. Similar habitats; local. ***Phlomis lychnitis*** is similar to *P. purpurea* but with narrow, **linear leaves** that are virtually stalkless below and stalkless above, and bright **yellow flowers**. Habitats similar to those of *P. purpurea*; frequent.

Prunella vulgaris **COMMON SELF-HEAL** A short to medium, slightly hairy, tufted perennial with stems ascending from a creeping base. Leaves lanceolate, toothed or not, stalked. Flowers blue-purple (rarely white), to 15 mm long, emerging from very dense, small, cylindrical heads of conspicuous purple, spiky calyces, with a pair of leaves immediately below. Damp and wooded places at higher altitude in the region; local.

Lamium amplexicaule **HENBIT DEADNETTLE** A short, scarcely branched, hairy annual. Leaves rounded or ovate; blunt-toothed and stalked below, unstalked and clasping the stem above. **Flowers held aloft**, pink-purple, to 20 mm long, with a slender, straight tube; calyx hairy. A common weed of cultivated ground.

1. *Sideritis linearifolia*
2. *Phlomis fruticosa*
3. *Phlomis purpurea*
4. *Prunella vulgaris*
5. *Stachys ocymastrum*
6. *Stachys germanica* PHOTO: TIM RICH

Stachys Annual or perennial herbs with flowers in dense, spike-like inflorescences; flowers with a tubular or bell-shaped calyx and 5 equal teeth; corolla 2-lipped with a flat or hooded upper lip and 3-lobed lower lip.

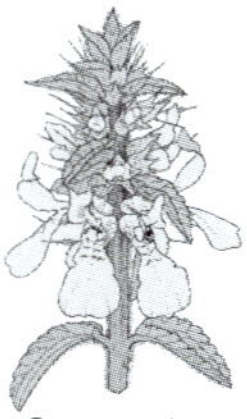

S. ocymastrum

Stachys ocymastrum An erect, pale green, hairy annual to 50 cm. Leaves to 50 mm long, oblong and pointed, slightly heart-shaped at the base, toothed and wavy along the margin. Flowers borne in dense whorls of 4–6 along the stem, congested above, laxer below; calyx densely hairy with 2 upper teeth as long as the tube and 3 shorter lower teeth; **corolla white**, shorter than the calyx, the upper lip entire. Common on the *matos*, especially near the coast. ***Stachys recta*** has **yellowish-white flowers** with purplish markings on the lower lip, borne in whorls of 6–16, crowded at the top. Dry, rocky places.

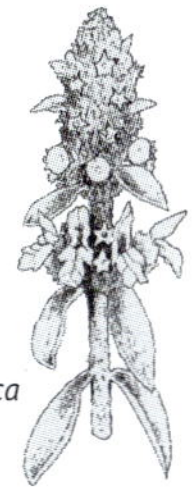

S. germanica

Stachys germanica DOWNY WOUNDWORT A **densely white-felted**, mint-like, erect perennial to 1 m. Leaves oblong and heart-shaped at the base, grey-woolly below and grey-green and less hairy above. Calyx with unequal teeth, the upper 2 less than half as long as the tube, but longer than the lower 3; corolla to 20 mm, **bright pink-purple**, borne in congested terminal inflorescences. Dry grassy places, fallow land and roadsides.

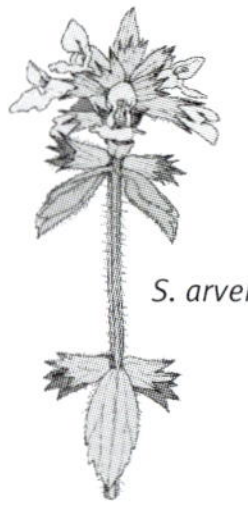

S. arvensis

Stachys arvensis **FIELD WOUNDWORT** A small, erect, hairy annual to 20 cm. Leaves ovate and heart-shaped at the base, hairy, toothed along the margin and wavy-edged. Whorls of 4–6 flowers, crowded above, distant below; calyx with teeth as long as the tube; corolla to 8 mm, white or pale pink and scarcely exceeding the calyx; upper lip entire. Cultivated land and sandy places.

C. nepeta

Calamintha nepeta **LESSER CALAMINT** A mint-like, medium, greyish, hairy perennial to 80 cm with creeping rhizomes. Stems erect and branched. Leaves ovate and shallowly toothed to untoothed; stalked. Flowers pale pink-purple with darker markings, borne in leafy whorls; calyx ribbed and purplish, with white **hairs protruding from the mouth**. Dry, fallow land and waste places.

C. vulgare

Clinopodium vulgare **WILD BASIL** A mint-like, short to medium, softly hairy, aromatic perennial with erect, branched or unbranched stems. Leaves ovate to lanceolate, slightly toothed and short-stalked. Flowers pink-purple, to 22 mm long, borne in distant whorls along an interrupted spike with prominent calyces; calyx green with purple ribs, more or less 2-lipped and prominently toothed. Disturbed, dry, grassy and stony habitats.

Micromeria graeca A slender, lax, **dwarf shrub** to 50 cm; woody at the base and variably hairy. Leaves small to 10 mm long, ovate below and linear-lanceolate above, with down-turned margins. Flowers borne in distant whorls forming long, slender, terminal inflorescences; corolla pink-purple; calyx woolly hairy in the throat with teeth almost as long as the tube and unequal. Dry, scrubby places.

1. *Stachys arvensis*
2. *Thymus camphoratus*
3. *Cleonia lusitanica* PHOTO: TIM RICH

Thymus **THYMES** Dwarf shrubs, woody at the base, and characteristically aromatic. Flowers borne in heads; calyx 2-lipped, and corolla 2-lipped with protruding stamens. Well represented in the region, but many species local.

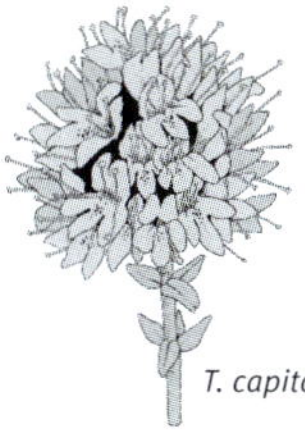

T. capitatus

Thymus capitatus A dense, cushion-like dwarf shrub to 30 cm with erect, woody branches with axillary leaf clusters. Leaves fleshy, **somewhat 3-angled**, to 10 mm long, stalkless, linear, more or less hairy and with inconspicuous lateral veins; gland-dotted, with a **flat margin**. Flowers pink, to 10 mm long, borne in dense, oblong terminal clusters with red-tinged bracts that overlap, forming cone-like heads; **calyx many-veined** (20–22). Dry, rocky places.

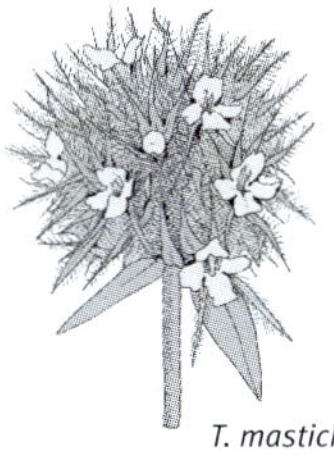

T. mastichina

Thymus mastichina A rather irregular dwarf shrub to 30 cm, not notably cushion-like or dense in profile with erect, woody branches. Leaves to 10 mm long, elliptical and wavy-margined, variably hairy. Inflorescence spherical; calyx to 6 mm with conspicuous, long, narrow, all similar teeth and **white flowers**; corolla scarcely exceeding the calyx. Sunny, exposed dry places. ***Thymus tomentosus*** is similar but with smaller leaves to only 5 mm long; **densely hairy**. Flowers borne in denser heads with smaller calyces (to 4 mm) with **shorter teeth** (to 1.5 mm). Habitats similar to those of *T. mastichina*.

T. camphoratus

Thymus cephalotus A dense dwarf shrub to 30 cm with erect, woody branches with axillary leaf clusters. Stem leaves to 10 mm long, linear and **densely hairy** with a few longer hairs, with down-turned margins. **Inflorescence large**, to 40 mm, oblong; bracts purplish, ovate and leathery. Calyx to 7 mm, cylindrical with lanceolate teeth. Corolla to 15 mm, purple. ***Thymus villosus*** is very similar (perhaps conspecific) but with bracts that are green below and **toothed or lobed**; corolla smaller, 6–10 mm.

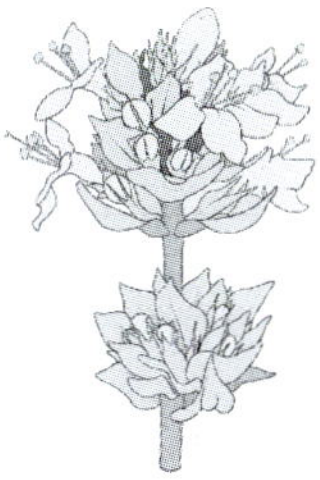

T. capitellatus

Thymus camphoratus A dense, aromatic (strawberry-scented), cushion-like dwarf shrub to 30 cm with erect, woody branches that have axillary leaf clusters. Leaves stalked, hairy and with down-turned margins. Inflorescence to 18 mm in diameter; bracts flushed with purple, noticeably so when in bud. Corolla pink-purple; calyx with thread-like teeth. Fixed sand dunes; very common around the coasts; the most common *Thymus* species in the region. ***Thymus capitellatus*** is very similar but with calyx teeth as long as wide (**not thread-like**), greenish (not purplish) bracts, and whitish flowers.

T. zygis

Thymus zygis A low to short dwarf shrub to 30 cm with erect, woody stems. Stem leaves to 10 mm, exceeding the axillary leaf clusters, linear, stalkless, hairy, with down-turned margins and fringed with hairs at the base. **Inflorescences forming long interrupted heads, to 10 cm**. Calyx greyish-green and bell-shaped with broad teeth; **corolla white**. *Matos*; dry rocky slopes.

Cleonia lusitanica A short, rather hairy plant with erect, purplish, solitary stems usually to just 10 cm tall. Leaves pinnately lobed below, grey-green and covered with short, white hairs. Flowers 2-lipped and tubular with rounded petal-lobes, violet (rarely yellow); flowers few, borne upwardly facing in leafy, terminal clusters. Low turf and scrubby habitats; local.

Origanum vulgare MARJORUM A medium, rather thyme-like, very aromatic perennial with erect, purplish stems. Leaves ovate and scarcely toothed or untoothed, short stalked or unstalked, and with leaf tufts in the axils. Flowers pink to reddish-purple, darker in bud, to 7 mm, borne in **broad, branched, panicle-like clusters**; calyx scarcely 2-lipped. Grassy places; infrequent.

Mentha MINTS Familiar aromatic perennial herbs with creeping rhizomes. Flowers cosexual or female in dense whorls; calyx actinomorphic or weakly 2-lipped; corolla weakly 2-lipped with 4 subequal lobes. Frequently hybridising.

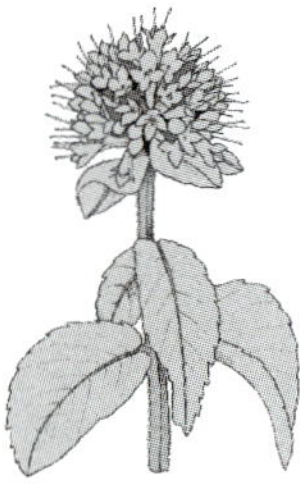
M. aquatica

Mentha aquatica WATER MINT A short to tall, vigorous and leafy, hairy to hairless perennial, strongly smelling of mint when crushed, with angled, purple stems. Leaves ovate, pointed, toothed and stalked. Flowers purple-white, to 6 mm long, borne in very dense, oblong heads with 1 or 2 whorls of flowers below; calyx veined and hairy. Always near water. ***Mentha pulegium*** is rather similar but with flowers forming long, interrupted spikes, **without a terminal head of flowers**; habitats similar to those of *M. aquatica*.

Mentha suaveolens APPLE MINT A rather small perennial to 40 cm with a sickly sweet scent and stems variably hairy. Leaves stalkless or virtually so, round-oblong and broadest near the base, to 45 mm long, toothed, hairy above and grey-hairy beneath. Flowers borne in many congested whorls forming long, dense inflorescences that are often branched; corolla pale pink or white; calyx hairy with subequal teeth. Cultivated as a culinary herb; grassy wasteland.

Lavandula LAVENDERS Familiar aromatic shrubs with narrow leaves and distinct bracts. Flowers borne in crowded, long-stalked terminal spikes; calyx with small teeth, 13-veined; corolla 2-lipped.

L. stoechas

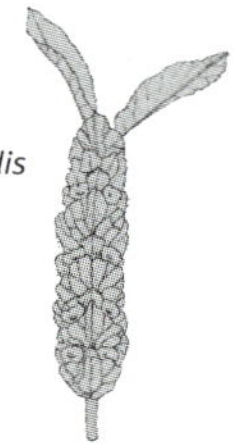
L. viridis

L. latifolia

Lavandula stoechas FRENCH/SPANISH LAVENDER A greyish shrub to 1 m with erect, much-branched stems. Leaves linear and untoothed. Flowers deep mauve, to 8 mm borne in short, dense spines, topped by **conspicuous, purple, flower-like bracts** to 50 mm long; inflorescences longer than their spikes. Very common on the *matos*, dry scrub, open woods and fixed dunes throughout the region. **Subsp. *lusitanica*** has **stalks longer than their short inflorescences**, and lower bracts not exceeding the calyx. Coastal sands. **Subsp. *pedunculata*** has stalks longer than their short inflorescences, and **lower bracts exceeding the calyx**.

Lavandula viridis GREEN LAVENDER Rather similar to *L. stoechas* but with **leaves and stems shortly hairy**, stalks longer than their inflorescences, and **white flowers with the bracts above the spike green**. Fairly common on the *matos*, but less frequent than *L. stoechas*.

Lavandula latifolia Rather like the commonly cultivated common lavender (*L. angustifolia*); a strongly aromatic (camphor-scented), grey-hairy shrub to 1 m with erect, much-branched stems. **Leaves densely** grey-hairy, linear and untoothed. Flowers small (to 10 mm), pale violet, borne on long-stalked spikes, **not topped by flowery bracts**. Bracts linear-lanceolate. *Matos* and dry rocky slopes.

L. multifida

Lavandula multifida CUT-LEAVED LAVENDER A grey-hairy, faintly aromatic shrub to 50 cm. Leaves green, sparsely hairy and **twice-pinnately lobed**. Flowers violet-blue, to 12 mm, borne in spikes up to 70 mm long; calyx 15-veined. Dry, stony, fallow land and low *matos*.

Rosmarinus officinalis ROSEMARY A familiar evergreen shrub to 2 m, characteristically aromatic (not strongly so in the Algarve). Branches brown and woody, erect to spreading. Leaves linear and leathery, mid to dark green, sharply pointed and with down-turned margins, needle-like. Flowers white, flushed pale purple, to 12 mm long, borne in small lateral clusters; corolla with 2 protruding stamens; calyx bell-shaped. Very common on the *matos*, dry scrub, open woods and fixed dunes throughout the region.

Salvia Herbs and shrubs with distinct whorls of (often purple) flowers forming a lax inflorescence. Both calyx and corolla 2-lipped, the upper corolla lip hooded, the lower 3-lobed; stamens 2, hinged in the middle, joined beneath the corolla hood.

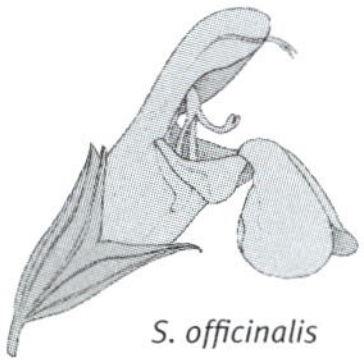

S. officinalis

Salvia officinalis SAGE A strongly aromatic, greyish shrub to 60 cm with erect branches that are woody beneath. Leaves broadly elliptical, greenish above and white-felted below, with a finely toothed margin. Flowers pale violet, blue or white, to 5 mm long; calyx flushed purple; bracts ovate and hairy. Cultivated and perhaps naturalised in the region.

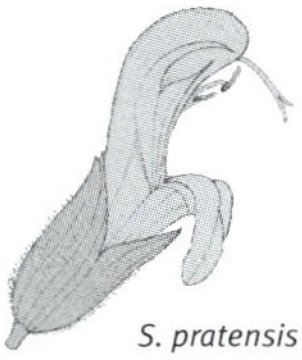

S. pratensis

Salvia pratensis FIELD SAGE A green perennial herb to 1 m, erect and branched, glandular above. Leaves long-stalked, ovate and heart-shaped at the base, crinkled along the margin, with spreading hairs above and more or less hairless beneath. Inflorescence lax and shortly branched. Whorls of 4–6 cosexual or female flowers, violet blue, **large (to 30 mm)**. Grassy places and open woods.

1. *Lavandula stoechas*
2. *Lavandula viridis*
3. *Rosmarinus officinalis*

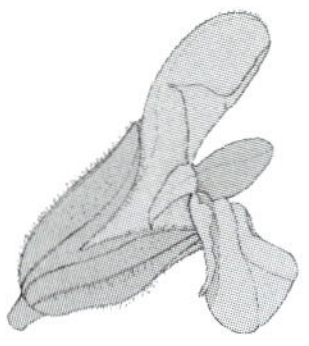

S. verbenaca

Salvia verbenaca WILD CLARY A short to tall perennial with erect stems that are glandular above. Basal leaves forming a rosette, **pinnately long-lobed**, wrinkled, toothed, long-stalked below, and short-stalked or stalkless above. Flowers pale blue or violet, up to 10 mm long, borne in lax whorls forming a spike; calyx bell-shaped and distinctly veined, enlarged in fruit. Fallow land, grassy places, roadsides and low coastal *matos*; common in the region.

OROBANCHACEAE

A distinctive family of hemiparasites and holoparasites that attach to the roots of their host plants as seedlings. Flower zygomorphic and usually more or less 2-lipped.

O. verna

Odontites verna A variable, short, hairy hemiparasitic annual to 50 cm. Stems erect, dark reddish-purple and slightly squared. Leaves opposite, lanceolate and normally toothed; bracts similar, exceeding the flowers. Flowers red-pink, 2-lipped, the lower deflexed, the anthers protruding. Grassy and sandy places, ruderal sites and roadsides. Local.

1. *Salvia verbenaca*
2. *Parentucellia viscosa*
3. *Bellardia trixago*
4. *Bellardia trixago* yellow form
5. *Cistanche phelypaea*
6. *Phelipanche ramosa*
7. *Phelipanche nana*

1

2

3

4

5

6

7

Odontites tenuifolia A densely pubescent hemiparasitic annual to 40 cm with linear leaves and dense, yellow flowers with protruding anthers. Heaths, pinewoods and maritime sandy places.

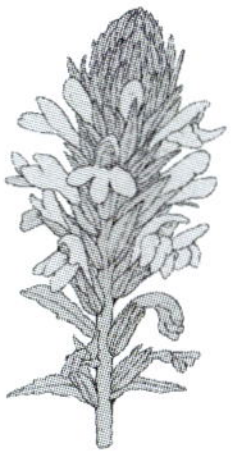

P. viscosa

Parentucellia viscosa A short, glandular-hairy hemiparasitic annual with erect, normally unbranched stems to 50 cm. Leaves opposite, lanceolate, deeply toothed and sessile; bracts similar, and decreasing in size along the stem. Flowers yellow (rarely white), 15–15 mm long; corolla 2-lipped, the upper lip hooded, the lower lip 3-lobed; calyx teeth linear-lanceolate, almost as long as the corolla. Damp grassy places near the sea; locally common. ***Parentucellia latifolia*** is similar but shorter (to 20 cm). **Leaves triangular-lanceolate and more deeply lobed**. Flowers small, 10–15 mm long, and pale purple (rarely yellow or white). Sandy and stony places.

P. latifolia

Bellardia trixago A short, glandular-hairy hemiparasitic annual with erect, simple stems to 50 cm. Leaves opposite, linear-lanceolate, toothed and sessile; bracts similar and decreasing in size along the stem. Flowers white, normally flushed with pink, or yellow, borne in a dense, **4-sided spike**. Ruderal sites, grassy places and *matos*; common.

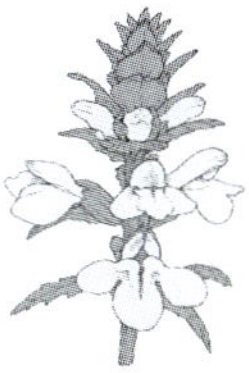

B. trixago

Cistanche phelypaea CISTANCHE A robust and imposing holoparasite with hairless, thick-stemmed inflorescences with ovate scale leaves. Flowers shiny, bright yellow borne in cone-shaped spikes; calyx 5-lobed; corolla with 5 more or less equal lobes. Parasitic on Amaranthaceae, especially *Arthrocnemum* spp. and *Atriplex halimus*; in maritime habitats. Often in large numbers when present, but local in coastal sands and estuaries only; locally common in Lagos, Alvor and on the western sea belt dune systems.

Phelipanche Holoparasites with simple or branched stems, often bluish, and with 2 bracteoles closely attached to the calyx. Flowers blue, white or cream, and constricted below the middle. Closely related to *Orobanche*, and still included in that genus by some authors.

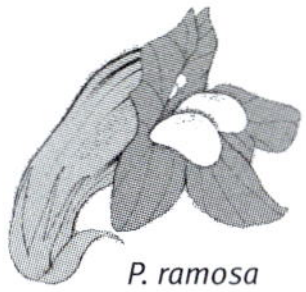

P. ramosa

P. nana

P. lavandulacea

Phelipanche ramosa BRANCHED BROOMRAPE A variable, short, bluish or purplish, glandular-pubescent annual with simple or branched stems, **usually** short (to 20 cm). **Flowers small (10–15 mm long)**, pale blue, violet or cream with a white patch at the base; stigmas white or pale blue; calyx teeth shorter than the corolla, **lower corolla teeth blunt**. On a wide variety of herbaceous hosts, often Solanaceae and Leguminosae. Sandy and disturbed habitats. ***Phelipanche nana*** is sometimes treated as a subspecies of *P. ramosa* and is a very similar, often yellowish or whitish annual with **pointed lower lobes** mid- to dark-blue flowers, and calyx teeth almost as long as the corolla. Parasitic on *Oxalis pes-caprae*; often confused with *P. ramosa* but probably the more common species in the region. There is a degree of continuous variation between *P. ramosa* and *P. nana*, and plants that are referable to both species often co-occur so their distinction (at least in the Algarve) is unclear. ***Phelipanche lavandulacea*** is similar to *P. ramosa* but **larger (to 50 cm)**, with **dense flowers** and a darker blue appearance. **Flowers to 25 mm long**; corolla white at the base, bright blue distally; stigma white, bluish or yellowish; filaments slightly hairy. On a variety of herbaceous hosts, often Leguminosae.

P. rosmarina

Disturbed, grassy places; infrequent. ***Phelipanche rosmarina*** is similar to *P. lavandulacea* but is **shorter (to 15 cm)**, with **a very constricted corolla**, virtually **hairless filaments** and a white stigma. Always on *Rosmarinus*. Local inland, but probably under-recorded in the region.

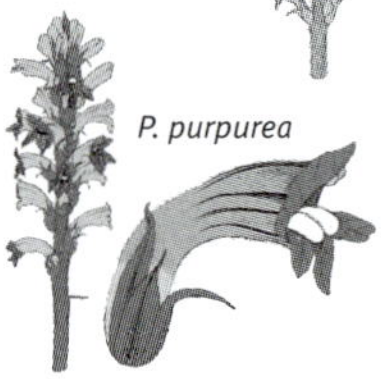
P. purpurea

Phelipanche purpurea Rather similar to *P. lavandulacea* but stems rarely branched, rather robust, to 30 cm. Inflorescence dense, with a dull purplish, **greyish or brownish** appearance. Corolla long, to 30 mm and distinctly constricted at the base; stigma white or pale blue. Parasitic on various hosts in the Asteraceae. Grassy places and *matos*. Fairly common inland but rather sporadic in appearance, and varying in abundance from year to year.

Orobanche A poorly understood genus of holoparasites with spikes of flowers that lack true leaves, roots and chlorophyll. Flowers have a distinctly 2-lipped corolla. Identification of the host is useful but problematic since plants might not parasitise the nearest green plant.

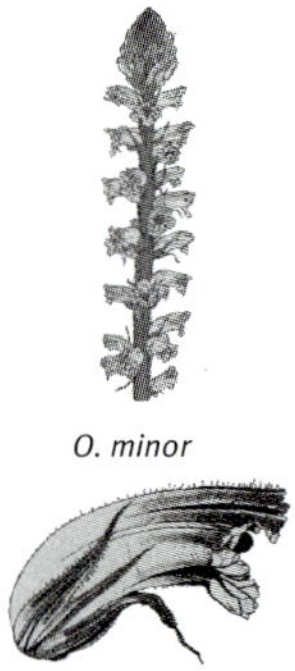
O. minor

Orobanche minor COMMON BROOMRAPE A variable species belonging to the taxonomically difficult *Minores* group. Usually with a rather lax inflorescence of **small flowers** (10–15 mm long) **with an evenly arched corolla**. Corolla pale yellow tinged and veined with violet; lower lip evenly lobed without a hairy margin, stigma usually pink. Bracts equalling or exceeding the corolla. Parasitic on a range of species but especially on legumes and composites, often on cultivated species in ruderal and sandy, disturbed habitats. Common throughout. Distinctive ecotypes in the *Minores* group occur in the region; for example, there are populations on *Carlina corymbosa* in the Cape St. Vincent area that have filiform bracts and calyx lobes, which may be conspecific with *O. minor* or the closely related species *O. picridis*; further systematic attention is required.

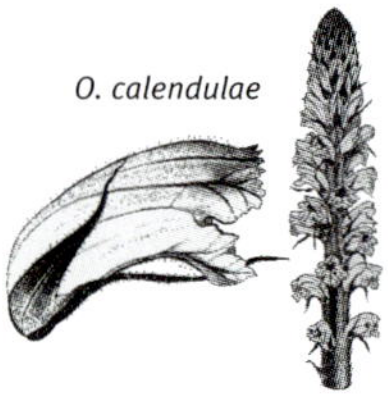
O. calendulae

Orobanche calendulae CALENDULA BROOMRAPE A poorly understood species. Similar to *O. minor* but with stems more robust and flowers over most of the stem. Flowers pale yellow tinged with violet or dull red; **stigma yellowish**. Parasitic on *Calendula suffruticosa* on maritime sea cliffs. Very local, in the Cape St. Vincent area; common in Castelejo.

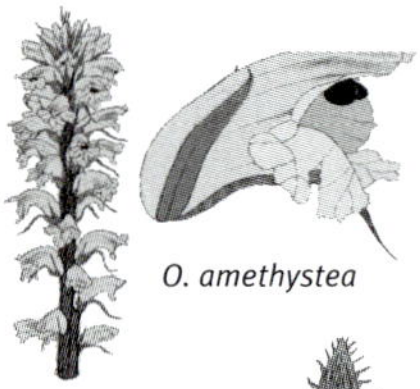
O. amethystea

Orobanche amethystea Like *O. minor* but generally more robust. **Flowers straight then abruptly arched**, pale yellow tinged and veined with violet (rarely all yellow or orange). Bracts exceeding the corolla, often markedly. Parasitic on various hosts, particularly composites and umbellifers, most commonly on *Eryngium campestre*. In grassy and sandy habitats, local. Distinctive ecotypes occur in the Cape St. Vincent area that may be conspecific to *O. amethystea* (or which may be a new species) and that are parasitic on *Plantago coronopus*. They have very dense inflorescences and robust stems. Further systematic attention is required.

O. rapum-genistae

Orobanche rapum-genistae GREATER BROOMRAPE Short to tall, **robust**, with glandular-hairy spikes with a brownish appearance. **Flowers large**, 20–25 mm long, dull brown, yellow or red, slightly foetid; lower corolla lip with a hairy margin, the **upper lip scarcely 2-lobed**; bracts exceeding the corolla, often markedly so; stigma yellow. Parasitic on shrubby legumes, especially *Cytisus* in rough grassland; local in the east of the region.

Orobanche gracilis **SLENDER BROOMRAPE** A short-medium, often robust (despite the name) species with reddish, lax spikes. Flowers large relative to the stem (5–15 mm), broadly campanulate, yellowish veined with red externally and **shiny red inside; bracts triangular, and shorter than the corolla**. Parasitic on legumes such as *Coronilla* spp. and *Cistus* spp., especially *C. crispus*. Locally common, in grassy places, scrub and woodland clearings at higher altitude or on *matos*.

1. *Orobanche minor*
2. *Orobanche calendulae*
3. *Orobanche amethystea*
4. *Orobanche* undescribed taxon parasitising *Carlina corymbosa*
5. *Orobanche rapum-genistae*
6. *Orobanche gracilis*

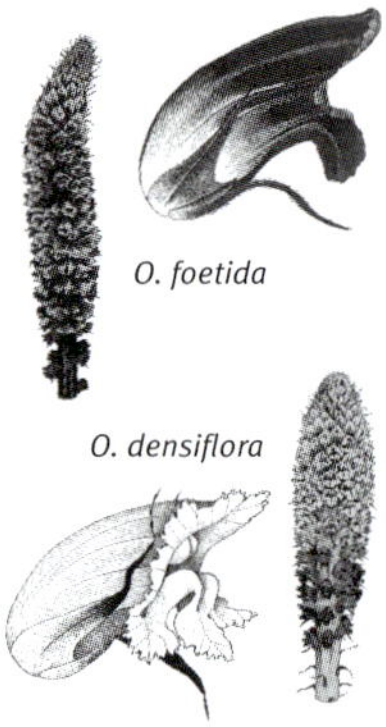

Orobanche foetida A tall, very robust species with densely flowered red or **blackish-red spikes**. Flowers purplish-red with a curved dorsal line and yellow or reddish stigma; slightly foetid; filaments inserted above the corolla base; scarcely hairy. Parasitic on legumes, especially *Ononis natrix* in disturbed habitats, road verges and dunes; common. Plant size seems to vary considerably with host. Bright yellow variants occur in some populations and have been erroneously confused with *O. densiflora*. ***Orobanche densiflora*** is hairy and has lemon yellow flowers with a **straight dorsal line**; corolla much exceeded by long, narrow bracts; **bracts dense and regular at the base of the stem**; and **filaments inserted at the base of the corolla**. Parasitic on *Lotus creticus* or sometimes *Medicago* spp. and *Ononis* spp. Apparently present in the Algarve (not observed there by the authors), but much confused with *O. foetida* and therefore possibly over-recorded or mistakenly identified in the region.

1

2

4

5

3

6

7

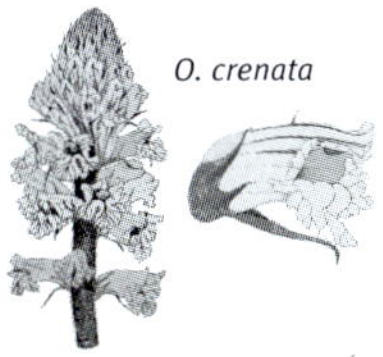
O. crenata

Orobanche crenata **BEAN BROOMRAPE** A short to medium annual with densely flowered spikes. Stems purplish. Flowers large (20–30 mm), broadly campanulate and **white** with violet veins, **fragrant**; stigma variable in colour. Parasitic on various legumes, particularly on cultivated peas and beans, sometimes in large numbers. Local, on cultivated and fallow ground.

PLANTAGINACEAE

A family that traditionally consisted of just three genera, *Bougueria*, *Littorella*, and *Plantago*, but which phylogenetic research has indicated is in fact allied with the Scrophulariaceae. The newly described Plantaginaceae includes numerous genera that were traditionally placed in the Scrophulariaceae, including toadflaxes (*Linaria* spp.), foxgloves (*Digitalis* spp.) and snap dragons (*Antirrhinum* spp.).

Herbs or shrubs with opposite to whorled, simple or compound leaves. Flower morphology variable but usually zygomorphic and 2-lipped. Fruit a capsule.

Plantago Small annual or perennial herbs with a basal rosette of leaves, opposite or alternate along the stem. Flowers small and inconspicuous, borne in dense heads or spikes; 4-parted; corolla papery; stamens 4 and protruding. Fruit a capsule.

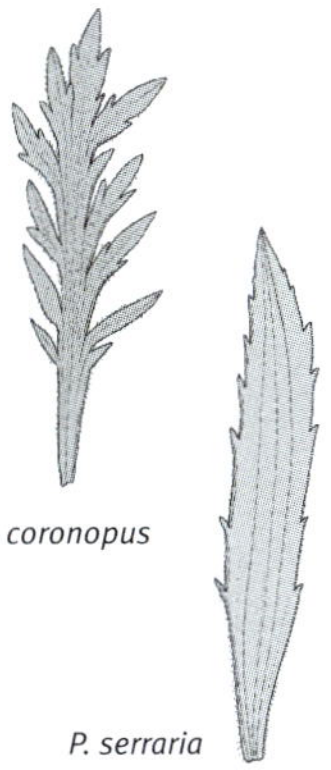
P. coronopus

P. serraria

Plantago coronopus **BUCK'S HORN PLANTAIN** A rather variable, low biennial or perennial with single or clustered leaf rosettes. Leaves linear to lanceolate, usually pinnately lobed, though sometimes unlobed, **not particularly fleshy**; hairless or finely hairy. Flowers yellowish-brown, to 3 mm, borne in spikes to 50 mm long, terminating from ungrooved, curved stems exceeding the leaves; anthers pale yellow. Very common in disturbed coastal habitats in the region, rarer further inland. ***Plantago serraria*** is similar to *P. coronopus* but with **leaves regularly toothed, not lobed,** often flushed purple. Flowering stems equalling or exceeding the leaves, with a **longer spike** of flowers to 10 cm long. Common in maritime rocky and sandy habitats or coastal *matos*. ***Plantago macrorhiza*** is similar to *P. coronopus* but always a perennial with a branched stock and multiple leaf rosettes. **Leaves fleshy** and rigid. Flower stalks few and exceeding the leaves, with spikes 20–70 mm long. Common in coastal habitats.

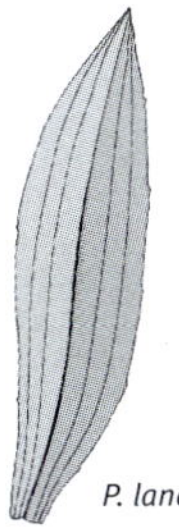
P. lanceolata

Plantago lanceolata **RIBWORT PLANTAIN** A variable low to medium, hairy or hairless perennial with several leaf rosettes. Leaves linear-lanceolate to lanceolate, toothed or untoothed, 3–5-veined, **strongly ribbed** and stalked. **Bracts hairless.** Flowers brown, to 4 mm, borne in short, blackish spikes on grooved stalks that markedly exceed the leaves; anthers pale yellow. A common weed on fallow land and grassy places throughout the region. ***Plantago lagopus*** is similar to *P. lanceolata* but smaller and **more hairy, especially the bracts**. Common in similar habitats to the previous species, especially near the coast.

1. *Orobanche foetida* common form
2. *Orobanche foetida* yellow form
3. *Orobanche crenata*
4. *Plantago coronopus*
5. *Plantago lanceolata*
6. *Plantago lagopus*
7. *Plantago serraria*

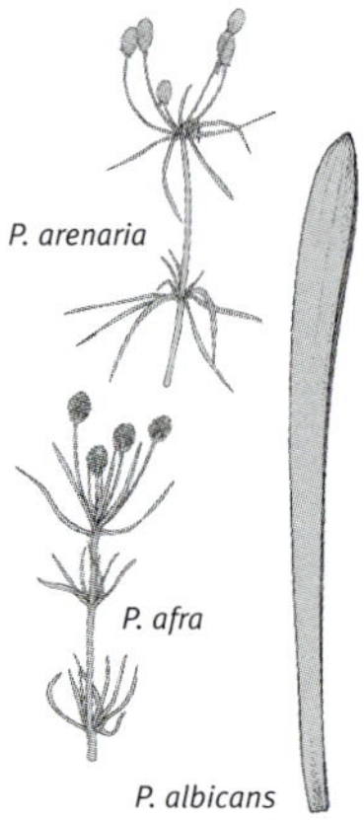

Plantago arenaria **BRANCHED PLANTAIN** A short to medium, hairy (but **not** markedly sticky) annual with **much-branched stems**, not typically plantain-like when in leaf. **Leaves linear** to linear-lanceolate, opposite or whorled, not fleshy and normally untoothed. Flowers brownish-white, to 4 mm, borne in round or conical spikes on spreading stalks; anthers pale yellow. Fairly common on coastal *matos* and in dry, sandy places. ***Plantago afra*** is very similar but extremely **sticky** and glandular-hairy above. On waste and fallow ground.

Plantago albicans **SILVERY PLANTAIN** A small, low to short, tufted, **silver-woolly** perennial with a woody stock. Leaves linear and often slightly twisted, 3-veined (obscured by hairs) and untoothed. Flowers greenish, borne in small, oblong spikes to 30 mm long on few, long, spreading stems; stamens not markedly protruding. Local on dry, bare ground, often coastal.

Antirrhinum Dwarf shrubs or woody-based herbs with zygomorphic, 'snap-dragon'-like flowers. Capsule unequally 2-parted.

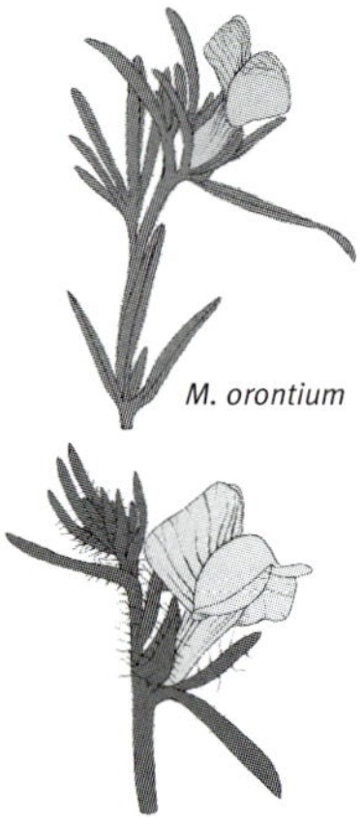

Antirrhinum majus **SNAPDRAGON** A medium, bushy perennial, to 1.5 m (usually less), much-branched below, with stems woody at the base. Leaves lanceolate to linear and wedge-shaped at the base, opposite or alternate. **Flowers snapdragon-like**: bright pink-purple (pale yellow in cultivated forms), large and showy, to 45 mm long. Very common in coastal areas on rocky slopes and fixed dunes. **Subsp. *cirrhigerum*** is **taller (to 2 m)** with **climbing stems** supported by other vegetation, and linear leaves widest at the middle. Common in the Cape St. Vincent area.

Misopates orontium **LESSER SNAPDRAGON** A short, little-branched, more or less hairless annual, to 30 cm. Leaves linear to elliptical, untoothed, opposite below and alternate above. Flowers pale pink or whitish, to 15 mm, and snapdragon-like; **calyx with long lobes** equalling the corolla. Capsule glandular-hairy. Local, probably often confused with *M. calycinum*. ***Misopates calycinum*** is similar but **larger (to 80 cm)**, with a dense raceme that elongates in fruit, and **larger, white flowers** (to 22 mm) that **exceed their calyces**. Common on waste ground near the sea and other sandy places in the region.

Linaria Annual or perennial herbs with simple, unstalked leaves, opposite or whorled, alternate above. Flowers in spikes or racemes, snapdragon-like but small; calyx unequally 5-lobed and short. Fruit a capsule. A well-represented genus, particularly in coastal districts of the region.

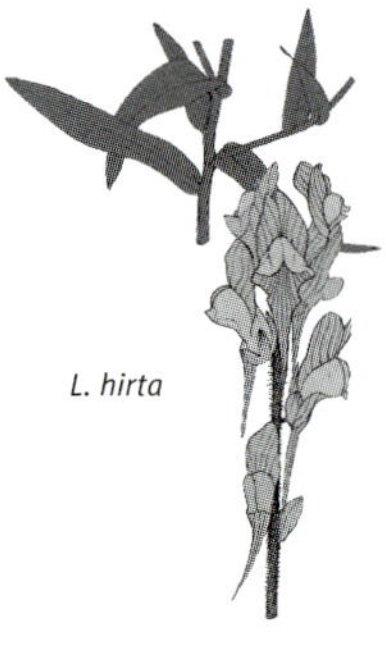

Linaria hirta A **densely glandular-hairy**, rather robust, unbranched annual to 50 cm tall. Leaves narrowly oblong-lanceolate, opposite below but **mostly alternate**, semi-clasping the stem at the base, and 3–4 times as long as wide. Flowers whitish or cream, to 30 mm long, with a reddish spur to 16 mm long. Local on coastal sands and sandy waste ground.

1. *Plantago afra*
2. *Antirrhinum majus*
3. *Misopates orontium*
4. *Misopates calycinum*
5. *Linaria ficalhoana*
6. *Linaria spartea*

Linaria ficalhoana A densely glandular-hairy, unbranched annual rather similar to *L. hirta* but **low to short** (to 15 cm tall, usually less). Leaves linear-oblong, to 7 mm long, and crowded on the stem, alternate above. Flowers bright yellow and spotted, to 18 mm long, with a reddish spur to 10 mm long. A local endemic on coastal sands, sometimes appearing in temporarily large numbers after fire damage.

Linaria pedunculata A rather glaucous, hairless annual to 30 cm, spreading to erect, stout, and branched above. Leaves oblong-lanceolate, fleshy, whorled below and alternate above. Racemes lax and leafy. **Flowers long-stalked**, stalks to 15 mm in flower and to 25 mm in fruit, held erect-spreading; corolla cream with violet veins, to 16 mm long, spur violet, to 7 mm long. Maritime sands.

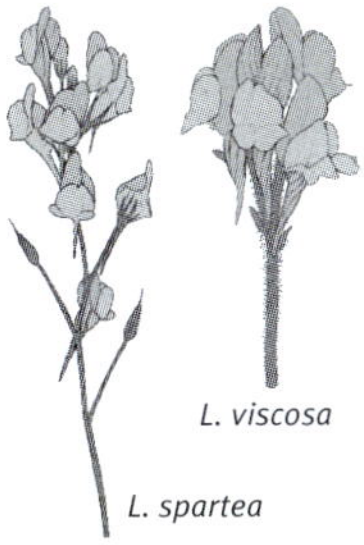

Linaria spartea A more or less hairless, erect annual to 40 cm. **Leaves small** (to 35 mm), linear, mostly alternate but whorled or opposite below; **distant**. Racemes short and lax in flower, even more so in fruit; flower stalks erect-spreading, 3–15 mm long, and **greatly exceeding their bracts**. Corolla bright yellow, to 30 mm long, with a yellow spur to 18 mm long; stigma distinctly 2-parted. Common in dry, sandy habitats. **Linaria viscosa** is similar to *L. spartea* but with denser leaves and the inflorescence **densely glandular-hairy** at least above. Flowers more crowded (not distant), borne on **short stalks** (to 8 mm). Habitats similar to those of *L. spartea*.

1

2

3

4

5

6

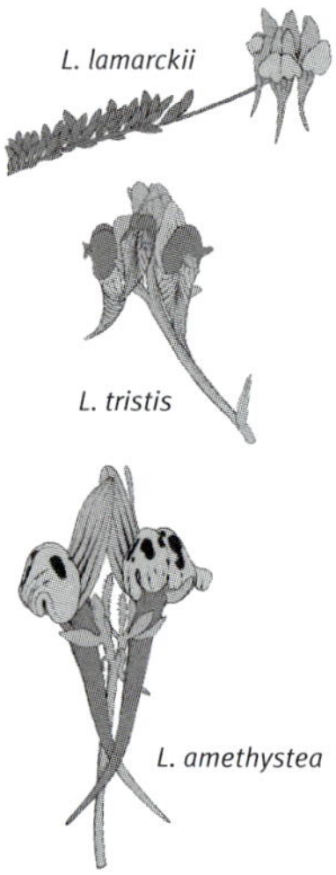

Linaria lamarckii A more or less hairless, **glaucous perennial** to 40 cm, usually unbranched and leafless at least for 20 mm below the lowermost flower; spreading, **not erect**. Leaves oblong to spatula-shaped, flat and somewhat fleshy, mostly alternate, at least above. Racemes many-flowered (to 12) and dense. Flowers short-stalked, to 5 mm in fruit, but **stalks always shorter than their bracts**; corolla deep yellow, to 30 mm long with a **long spur** to 17 mm long, striped orange-brown. Local on maritime sands. ***Linaria tristis*** is similar to *L. lamarckii* but with a densely glandular-hairy inflorescence and flowers with a **short spur** (11–13 mm). Limestone rocks.

Linaria amethystea A short, **erect**, usually unbranched annual to less than 35 cm, hairless below, but hairy (with **violet hairs**) above. Leaves linear to lanceolate, at least 3 times as long as wide, whorled below, alternate above. Racemes few-flowered and lax with up to 10 flowers (usually fewer). Flowers borne in short stalks, not exceeding their bracts; corolla usually blue-violet or whitish and spotted (yellow and cream forms exist), to 27 mm long, spur to 15 mm and slender. Common in open, sunny places, often coastal.

1. *Linaria amethystea*
2. *Linaria amethystea*
3. *Linaria algarviana*
4. *Cymbalaria muralis*
5. *Digitalis purpurea*
6. *Eryngium maritimum*

L. algarviana

Linaria algarviana A small annual rather similar to *L. amethystea* but **smaller**, with **prostrate to ascending** stems to 30 cm (usually less), hairy above, hairless below. Leaves few, to 8 mm long, linear and alternate. **Racemes lax with few flowers** (up to 8, usually fewer). Flowers borne on stalks longer than their bracts; corolla to 25 mm, violet, spotted with white or yellow; lips of the corolla more or less touching (**not divergent**); spur to 12 mm; stigma deeply 2-parted. Seeds black and wrinkled. A rare endemic of coastal shales and sands in the Cape St. Vincent area.

L. micrantha

Linaria micrantha A leafy, slightly glaucous annual, hairless throughout except on the inflorescence; erect to 30 cm. Leaves linear to lanceolate. Racemes dense in fruit. Flowers borne in rather inconspicuous, small terminal clusters; **very small** (to 9 mm long), lilac, with a **minute spur** that is less than 1 mm long and straight or curved. Cultivated and waste places.

Cymbalaria muralis **IVY-LEAVED TOADFLAX** A trailing, purplish, tufted, hairless perennial. Leaves alternate, rounded with 5–9 lobes, long-stalked. Flowers small, lilac with a yellowish palate, to 15 mm long, borne on long, slender stalks, spur to 3 mm long. Capsule hairless. Local in damp shady places on rocks on higher ground.

Digitalis purpurea **FOXGLOVE** A familiar biennial or perennial herb with soft, ovate to lanceolate basal leaves, rather hairy and long-stalked. Flowers showy, borne in long, robust racemes that are many-flowered and unbranched; corolla large and bell-shaped, to 55 mm long, pink-purple and spotted internally. Locally common in damp, shady woods on higher ground.

V. anagallis-aquatica

Veronica anagallis-aquatica **BLUE WATER SPEEDWELL** A medium, hairless or slightly hairy perennial with erect, branched or unbranched stems. Leaves opposite, ovate to lanceolate and scarcely toothed, stalked below, unstalked and semi-clasping the stem above. Flowers blue with violet veins, to 10 mm, borne in slender, paired racemes. Capsules rounded or elliptical, hairless. Frequent in damp, seasonally flooded habitats or by streams.

Order: APIALES

APIACEAE

A very large and important family, well represented in the region. Mostly herbs with alternate leaves, often pinnately divided with sheath-like bases and characteristically carrot-like. Flowers borne in normally compound umbels very typical of the family. Flowers typically cosexual, small, often green, yellow or white and 5-parted; petals free; carpels 2, fused and each ending in a style, often winged. Fruit morphology is often an important species diagnostic.

Eryngium Hairless, often distinctly spiny herbs with dense flowers borne in rounded, thistle-like heads, surrounded by spiny bracts. Fruit rounded and often scaly.

E. maritimum

Eryngium maritimum **SEA HOLLY** A short, rigid perennial with leathery, **bluish, white-veined leaves**, 3–5-lobed with an undulate, spiny margin. Flowers pale blue, borne in dense, rounded heads to 30 mm framed by coarsely spiny bracts; upper stems flushed bright blue. Very common on coastal sands and dunes. ***Eryngium dilatatum*** is rather similar but with 3- to **pinnately lobed lower leaves** that are softly spine-toothed, and with **winged stalks**. Flower-heads to 15 mm, framed by bracts with 3–6 pairs of spines. Dry habitats inland; local.

E. campestre

Eryngium campestre **FIELD ERYNGO** A short, rigid perennial with green or yellowish, leathery, 3-lobed basal leaves that are spine-toothed with unwinged stems; stem leaves unstalked and clasping the stem. Flowers pale green-white or yellowish, borne in dense rounded heads (to 15 mm). Bracts with or without a single pair of spines. Common in dry, sandy or grassy places both coastally and inland. ***Eryngium corniculatum*** is similar but with **undivided basal leaves** with **swollen, segmented stalks**, and a bluish flower-head, the axis of which projects into a **bract-like appendage**. Seasonally flooded habitats.

L. cuminoides

Lagoecia cuminoides **LAGOECIA** An annual with pinnate leaves with toothed segments; upper leaves yarrow-like, deeply divided and with narrow lobes. Umbels rather feathery, white, rounded and dense, to 15 mm. Fruit to 2 mm with brittle hairs. Local in dry, fallow habitats.

Smyrnium Biennial herbs in which the 2–3 leaf lobes arise from a single point (ternate). Umbels yellow, the flowers lacking sepals. Fruit egg-shaped with slender ridges and oil glands.

S. olusatrum

Smyrnium olusatrum **ALEXANDERS** A tall, pungent, hairless biennial to 1.5 m. Stem robust, hollow when mature, bearing once to twice pinnate leaves with triangular, toothed lobes, shiny green below, yellow above, borne on short, somewhat inflated stalks. Umbels terminal and borne in the leaf axils with 7–15 rays. Flowers yellow, to 3 mm across without sepals. Fruit ovoid, to 8 mm and **black** when ripe. Common in damp, coastal habitats.

Smyrnium perfoliatum **PERFOLIATE ALEXANDERS** A virtually hairless, tall biennial to 1.5 m. Stem robust, angled with longitudinal narrow wings, solid and hairy at the nodes. Basal leaves with 1–2 lobes that are oval and toothed; upper leaves yellowish, simple, oval and markedly **clasping the stem** with heart-shaped bases. Umbels yellow, rounded with 5–12 rays and no bracts. Fruit broader than long, brown-black. Wooded and open grassy slopes; local.

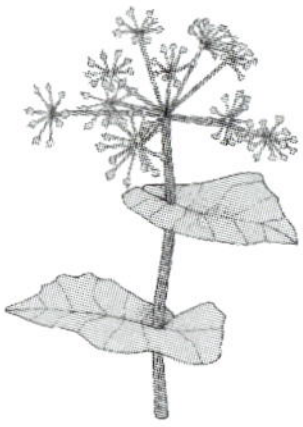

S. perfoliatum

Bunium pachypodium A short to medium perennial with leaves 2–3-times pinnate with **linear**, blunt-tipped lobes, withered at the base when mature. Umbels white, lax, with up to 15 spreading rays. Bracts 6–8. Fruit to 6 mm long, with thick ridges. Cultivated land.

Bunium bulbocastanum **GREAT PIGNUT** A large, erect perennial to 1 m. Stems remain solid (not hollow) after flowering. Basal leaves twice to 3-times pinnate, to 15 cm long. Umbels to 80 mm across with up to 10 bracts and as many bracteoles. Styles bent in fruit. Local on roadsides and in grassy habitats.

1. *Eryngium campestre*
2. *Eryngium campestre*
3. *Smyrnium olusatrum*
4. *Smyrnium olusatrum*
5. *Scandix pectin-veneris*

Scandix Slender annual herbs with thread-like, feathery leaves, white flowers without sepals, and conspicuous long, needle-like fruits with cylindrical beaks.

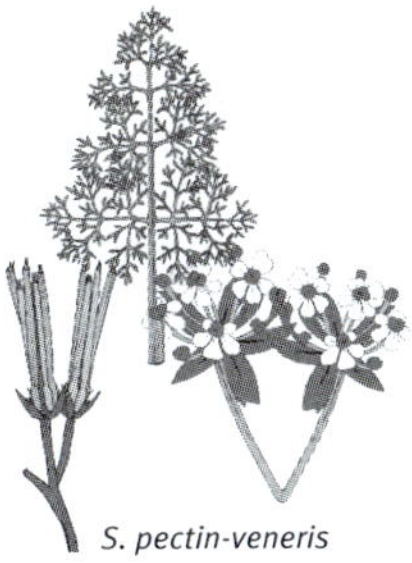

S. pectin-veneris

E. maritimum

Scandix pectin-veneris SHEPHERD'S NEEDLE A small, spreading, hairless annual to 30 cm with leaves 2–3-times pinnate, the segments widened towards the tips. Umbels simple or with 2–3 rays. Flowers small and white. Fruits borne in claw-like clusters and highly distinctive: **elongated to 80 mm long,** at least half the length comprising the seedless beak, which is **flattened and distinct**. Very common on disturbed and fallow ground. ***Scandix australis*** is very similar but the fruit is **scarcely compressed**, and the beak **indistinct** from the rest of the fruit. Pine forests and shady places.

Conopodium capillfolium A small, tuberous plant to 80 cm. Leaves ovate, 2–3-times pinnately divided at the base; stem leaves with long, **linear** segments; sheaths of the middle and upper leaves 7 mm or more, at least ¼ the length of lamina. Umbels with 6–20 rays and 0–2 bracts, and 5 or more secondary bracts (bracteoles). Petals white with a brown vein on the back. Fruit oblong, laterally flattened. Grassy places. ***Conopodium bourgaei*** is very similar but with broader linear-lanceolate lobes on the stem leaves, no bracts, and 0–1 bracteoles.

1
2
3
4
5
6

Crithmum Leaves 1–2-times pinnate and fleshy. Petals yellow-green. Fruit oblong, not compressed and rather corky.

C. maritimum

Crithmum maritimum ROCK SAMPHIRE A short to medium, somewhat bushy, greyish, hairless perennial; woody at the base. Leaves 1–2-times pinnate with almost cylindrical, untoothed **fleshy** segments, the base membranous and clasping the stem. Flowers yellow-green, in umbels to 60 mm across with 8–36 rays, sepals absent. Bracts and secondary bracts numerous, triangular. Fruit to 6 mm, oval and corky, later purple. Common on cliff tops, rocky shores and other maritime environments.

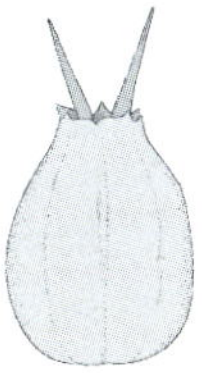
O. globulosa

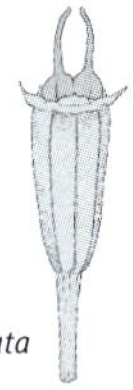
O. crocata

Oenanthe globulosa MEDITERRANEAN WATER DROPWORT A **short**, much-branched, medium perennial with egg-shaped tubers to 40 cm. Stems rather bluish-grey, hollow and distinctly **grooved**. Basal leaves 2-times pinnate with ovate to linear lobes, the upper leaves less divided and narrower. Umbels white with 3–16 (normally less than 10) lax, disparate rays. Flowers male or cosexual, sepals persistent, **flower stalks thickened in fruit**. Fruit rounded, to 5 mm with thickened lateral ridges. Locally common in marshes close to the sea. ***Oenanthe crocata*** is similar but **taller** and more robust (to 1.5 m), with **flower stalks not thickened in fruit**, and the **fruits cylindrical**. More frequent than *O. globulosa* but in similar habitats.

Foeniculum Leaves feathery, 3–4-times pinnate, aromatic. Sepals absent, petals yellow. Fruit distinctly ridged and scarcely compressed.

F. vulgare

Foeniculum vulgare FENNEL A robust, tall, hairless, **strongly aromatic** perennial to 2 m. Stems bluish, hollow when mature. Leaves feathery, 3–4-times pinnate with **thread-like segments**, light green and with sheathing bases. Umbels yellow with up to 30 rays without any bracts, flowers borne in mid summer. Fruit oblong, to 10 mm long, ridged. Very common on the *matos*.

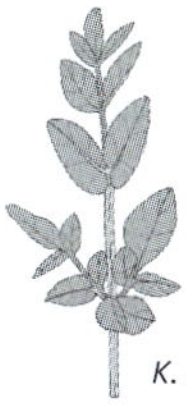
K. sicula

Kundmannia sicula KUNDMANNIA A medium, hairless perennial with erect stems. Lower leaves twice pinnate with an extra pair of lobes at the base of each main lobe; lobes ovate, toothed, upper leaves once pinnate and coarsely toothed. Umbels yellow with up to 30 rays, bracts numerous and linear and keeled backwards. Fruit cylindrical to 10 mm long with slender ridges. Dry slopes and fallow ground; uncommon.

Cachrys Leaves 2–4-times pinnate with linear lobes. Flowers yellow, fruit with undulate wings and ridges.

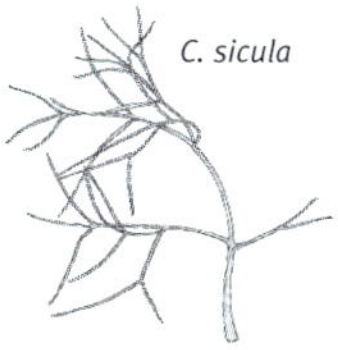
C. sicula

Cachrys sicula A bushy, bluish-green perennial to 50 cm with a solid, grooved stem. Leaves fennel-like; 2–3-times pinnate, ovate in outline with linear, somewhat **flopping lobes** to 50 mm long that are rough along the margins. Flowers with up to 30 rays, yellow, and with sepals present. Bracts of the terminal umbel 1–2-times pinnately lobed. Fruit to 15 mm, wedge-shaped and with prominent ridges. Dry slopes and *matos*.

---

1. *Crithmum maritimum*
2. *Crithmum maritimum*
3. *Oenanthe globulosa*
4. *Oenanthe crocata*
5. *Foeniculum vulgare*
6. *Foeniculum vulgare*

C. libanotis

Cachrys libanotis Similar to *C. sicula* but smaller and stouter, the leaves less divided (2–3-times pinnate) with **rigid, slightly toothed lobes** to 10 mm long. Flowers with 8–15 rays with simple or pinnate bracts. Fruits less than 10 mm, with indistinct ridges, smooth or with flattened spines. Common on sea cliffs near Sagres and on Cape St. Vincent (populations there have stunted, lanceolate not linear leaf segments, superficially similar to those of co-occurring *Ferula tingitana*).

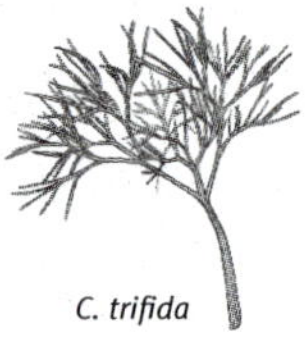

C. trifida

Cachrys trifida A large, hairless, light blue-green perennial to 1 m tall. Leaves triangular in outline, to 40 cm long, **4–7-times pinnate** and somewhat feathery with linear lobes to 30 mm long. Umbels yellow with 10–20 rays and with few or no bracts. Flowers without sepals. Fruit to 15 mm with wide, smooth ridges. Very common on the *matos* and in coastal habitats throughout the region.

Bupleurum Hairless annuals or perennials with simple, undivided leaves (atypical of the family). Flowers pale green or yellow, borne in small umbels surrounded by petal-like bracts, sepals absent, petals not notched. Fruit prominently ridged.

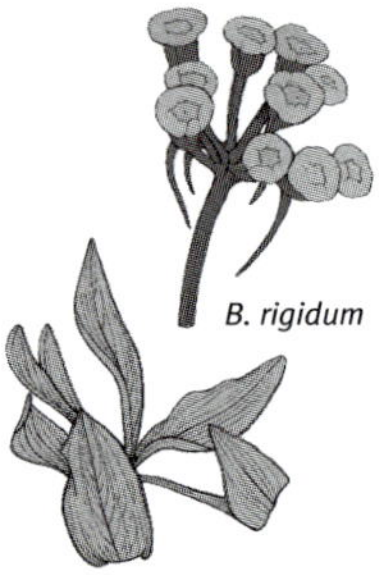

B. rigidum

Bupleurum rigidum A perennial to 1.5 m, woody at the base with leathery leaves that are stalked and clasping the stem, with 3–11 veins. Branches numerous. Umbels with 2–5 rays and short, slender bracts. Fruit to 4 mm long, with prominent ridges. Common on dry rocky habitats and sea cliffs. ***Bupleurum acutifolium*** has narrow leaves with less conspicuous veins, those from the upper stems much longer than the lower. Recorded from north of the Algarve near Odemira, and possibly present further south. ***Bupleurum fruticosum*** is an aromatic, evergreen **shrub**. *Matos* and roadside scrub.

Apium nodiflorum FOOL'S WATERCRESS A medium to tall, yellowish biennial to 1 m with a celery-like smell. Stems hollow, spreading and rooting at the nodes, then ascending. Leaves shiny, once pinnate with ovate to lanceolate segments to 60 mm long. Flowers with whitish petals and 3–12 rays. Fruit ovoid and longer than wide. Wet places, ditches and riverbanks.

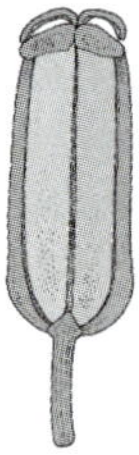

R. segetum

Ridolfia segetum RIDOLFIA A fennel-like, bluish-green annual, unpleasant-smelling when crushed. Leaves 4-times pinnate with thread-like lobes and inflated petioles. Umbels yellow with many (to 60) slender, curved rays and no bracts or sepals. Fruit small, to 2.5 mm long with slender ridges. Cultivated and waste land.

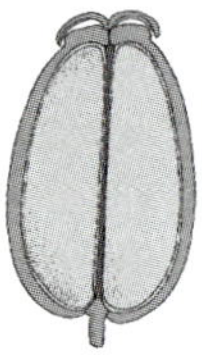

A. visnaga

Ammi visnaga A robust, carrot-like annual or biennial to 1 m. Lower leaves once pinnate, upper leaves further divided, all somewhat feathery with narrow to thread-like lobes. Inflorescence superficially similar to *Daucus* spp.; flowers white; rays numerous (to 150), slender and spreading in flower but much thickened and erect in fruit. Bracts divided, equalling or exceeding the rays. A weed of rocky and scrubby waste ground.

Ammoides pusilla A slender, carrot-like, hairless annual to 50 cm or less. Leaves bluish, twice pinnate below with 7–11 pairs of very short lobes, the upper leaves with longer, **linear segments**. Umbels white; with 5–11 unequal rays; bracts absent or few, **sepals conspicuous**. Fruit small to 1 mm, with slender ridges. Dry habitats.

Capnophyllum peregrinum A small, carrot-like, hairless annual to 50 cm or less. Stems solid and grooved. Leaves broadly triangular in outline, 3-times pinnate with linear segments. Flowers white, rays 2–5. Bracts absent, bracteoles 4–6, shortly triangular. Fruit with very prominent ridges, ovoid and laterally compressed, to 6 mm.

Ferula Perennials with leaves 3–4-times pinnate and linear lobes; somewhat fennel-like. Flowers yellow. Fruits strongly compressed dorsally.

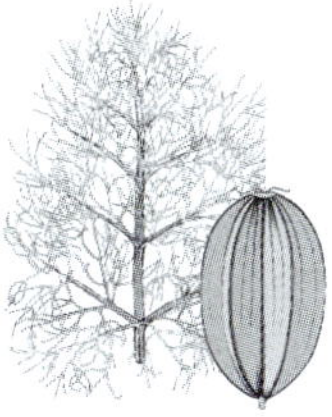

F. communis

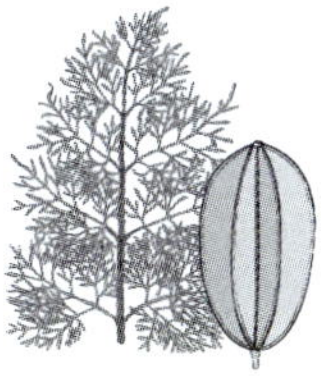

F. tingitana

Ferula communis GIANT FENNEL A giant, robust perennial herb to 3 m, bushy at the base and a conspicuous feature of the landscape when in flower, and long-persisting when dead. Leaves 3–4-times pinnate with bright green, **thread-like segments to 50 mm long**, and prominent sheathing bases. Upper leaves progressively reduced to sheathing bases only. Umbels large and rounded, bright yellow-green, with 20–40 rays. Bracts absent and bracteoles few and soon-falling. Fruit large to 15 mm long, elliptical, flattened with thin dorsal wings and numerous resin canals. Commonly recorded in the region, but probably confused with *F. tingitana*, therefore the true distribution (if present at all in the region) is unclear. ***Ferula tingitana*** is similar and often confused with *F. communis*. It is less robust with darker green leaves with **shorter leaf lobes to 10 mm long**, with or without down-turned margins, the ultimate lobes partly united to one another; similar to *F. communis* when in flower and in fruit. Common in rocky, scrubby and wind-swept *matos* and on roadsides. Large stands occur on Cape St. Vincent.

1. *Cachrys libanotis*
2. *Cachrys trifida*
3. *Cachrys trifida*
4. *Ferula tingitana*
5. *Ferula tingitana*
6. *Ferula tingitana*

1
2
3
4
5

E. foetidum

Ferulago capillaris A bushy, light green perennial to 2 m or less with weakly grooved stems. Leaves fennel-like and triangular in outline, 2–3-times pinnate with short, linear, **thread-like lobes**. **Bracts and bracteoles well-developed and conspicuous**. Sepals minute, petals yellow. Fruit narrowly ovate with well-developed lateral wings. Recorded near Tavira, but the distribution in the Algarve is poorly known.

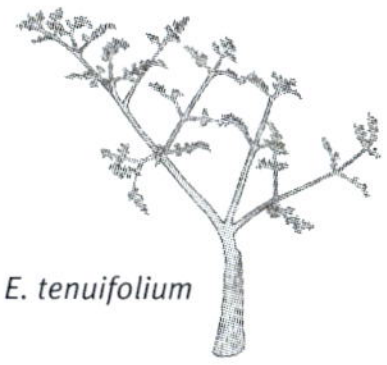

E. tenuifolium

Elaeoselinum foetidum A bright green, more or less hairless, rather fennel-like perennial with leaves 3-times pinnate with lobes to 10 mm long. Flowers bright yellow with 10–20 rays, 0–1 bracts, and numerous bracteoles. Fruits with **wide lateral wings that extend beyond the top of the fruit**. Dry sunny slopes and *matos*. Similar to, and confused with, *Thapsia villosa* but with bushier (not flat) rosettes with less conspicuous petioles, and later to flower (April–May). **_Elaeoselinum tenuifolium_** is similar but with small, sparse, lanceolate leaf lobes, and **numerous bracts** and bracteoles. Local.

T. villosa var. *villosa*

Thapsia villosa var. _villosa_ A robust, dark green perennial to 2 m. Leaves flat, triangular, to 30 cm long and forming a rosette with conspicuous, often reddish sheathing leaf bases. Leaves 3–4-times pinnate, the ultimate lobes toothed, and bristly hairy along the centre and on the lobes. Umbels yellow, borne on erect stems, with 9–24 rays and very few bracts. Fruit to 15 mm, elliptical with lateral wings to 3 mm wide, and deeply notched at both ends. Common on the *matos*. **_Thapsia transtagana_** (previously confused with the Eastern European species *T. garganica*) is similar in form to *T. villosa* but produces a larger rosette of rather bushy (not flat) leaves that are blue-green beneath, with longer, narrower, unlobed segments. Local. **_Thapsia maxima_** is similar to *T. villosa* but often larger and more robust, with dark green, rather shiny leaves that are **pinnately divided once only** with toothed or shallowly lobed segments. Fairly common on roadsides, sea cliffs and dunes.

T. transtagana

Prangos trifida A hairless, herbaceous, very bushy, light yellow-green, fennel-like perennial. Leaves roughly triangular in outline, **6 times divided with long, linear lobes**, giving the plant a feathery appearance; aromatic when crushed. Inflorescences stout with umbels of up to 15 rays of bright yellow flowers. **Fruits large, unwinged, to 20 mm long, ovoid, purplish and rather corky when ripe.** Very common on undisturbed matos.

T. maxima

Margotia gummifera A herbaceous perennial with a flat, basal rosette of leaves borne on purplish stems. Leaves flat, 3 to 4 times pinnate, the terminal divisions linear to linear-lanceolate; dark shiny green or purplish, and **very strongly aromatic** when crushed. Inflorescences to 1 m tall with 3–8 linear to linear-lanceolate, drooping bracts. Flowers white with 5 incurved petals. Seeds large, winged and papery – similar to those of *Thapsia* spp. Very common on coastal cliffs, dunes and matos in the region.

1. *Thapsia villosa*
2. *Thapsia maxima*
3. *Prangos trifida*
4. *Margotia gummifera*
5. *Margotia gummifera*

Distichoselinum tenuifolium A herbaceous fennel-like, hairless, bushy perennial. Leaves bushy, 4–5 times pinnate with small, linear, pointed terminal segments; rather shiny mid to dark green and aromatic when crushed. Inflorescences to 2 m tall (often less) with 8–12 reflexed bracts and umbels of yellow flowers. Seeds large, winged and papery. Very common on coastal cliffs, dunes and matos in the region.

P. pumila

Torilis nodosa KNOTTED HEDGE PARSLEY A low to short, **prostrate**, rough-hairy annual. Leaves once or twice pinnate with deeply toothed segments. Flowers pinkish-white, clustered on stalkless or short-stalked leaf-opposed umbels with 2–3 very short rays. Fruit egg-shaped, to 2 mm long with warts and straight bristles. Common on open *matos* or waste ground.

Pseudorlaya pumila PSEUDORLAYA A **low to short, densely hairy**, rather fleshy annual, branched from the base. Leaves 2–3-times pinnate with oval segments. Umbels white to pale purple with 2–5 unequal rays. Petals more or less equal, some larger at the perimeter. Bracts 2–5, linear. Fruit elliptical, to 10 mm long, ridged and with hooked spines. Common on coastal sands, dunes, and sandy cliff tops.

Daucus Annual or biennial carrots with leaves 2–3-times pinnate. Umbels white, with pinnately lobed bracts; petals unequal, those of the outer flowers often larger. Fruits elliptical and spiny.

Daucus carota WILD CARROT A very variable, short to tall, hairy or hairless annual or biennial with solid, often ridged stems. Leaves feathery with linear to lanceolate segments. Umbels white, many-rayed, often with a single purple flower in the centre, or purplish throughout. **Bracts pinnately lobed**. Fruit oblong, to 4 mm and shortly spiny; the spines not distinctly fused at the base; the umbel rays becoming conspicuously incurved when dry. Common in grassy places throughout the region. **Subsp. *carota*** grows to 1 m, with basal leaves 1–4 times divided (pinnatisect) and umbels up to 11 cm in diameter when in bloom. **Subsp. *maxima*** grows to 2 m, with basal leaves 1–3 times pinnatisect, and **large** umbels 12–23 cm across.

Daucus muricatus is somewhat similar but with umbels with markedly unequal rays, and fruits with slivery-white rows of long spines that are joined at the base.

Daucus halophilus A short, robust, rather **fleshy**, sparsely branched biennial to 30 cm. Leaves twice pinnate with rather **succulent**, ovate, more or less entire lobes. Umbels with 12–40 long rays. Bracts shorter than the rays with 5–7 lanceolate lobes; bracteoles 3-lobed at the tip. Flowers white, pinkish at the centre of the umbel, with unequally large petals on the perimeter. Fruit small, 2–3 mm with yellowish spines joined at the base forming a crest. Rocky shores and maritime cliffs in the Cape St. Vincent area.

1. *Distichoselinum tenuifolium*
2. *Distichoselinum tenuifolium*
3. *Daucus carota*
4. *Daucus halophilus*
5. *Pittosporum tobira*

PITTOSPORACEAE

A family of trees, shrubs and lianas with simple, alternate leaves and 5-parted flowers.

Pittosporum tobira A dense evergreen shrub to 5 m. Leaves alternate, ovate, deep green and leathery with inrolled margins. Flowers cream, to 22 mm across with 5 petals, and fragrant, borne in dense, flat-topped clusters. Widely cultivated in towns and on roadsides in the Algarve. ***Pittosporum undulatum*** has broad, mid-green leaves that are narrowed at the base, with **undulate (not inrolled) margins**. Occasionally planted in the region.

Order: **DIPSACALES**

ADOXACEAE

A small family of plants with opposite, toothed leaves and cymose inflorescences; previously grouped with the family Caprifoliaceae.

Vibernum tinus An evergreen shrub to 7 m tall with more or less hairless stems, and dark green, ovate, short-stalked leaves that are sparsely hairy beneath. Flowers white, pinkish in bud, borne in dense, flattened heads to 90 mm across. Fruit a berry to 7 mm, **blue** when ripe. Locally common in hill scrub and cork oak forests.

CAPRIFOLIACEAE

Woody shrubs and climbers with opposite leaves. Flowers 5-parted and cosexual, solitary, paired or in showy panicles on shrubs; calyx small; corolla actinomorphic or 2-lipped, fused below to form a tube; stamens mostly 5; ovary inferior. Fruit a berry or nutlet.

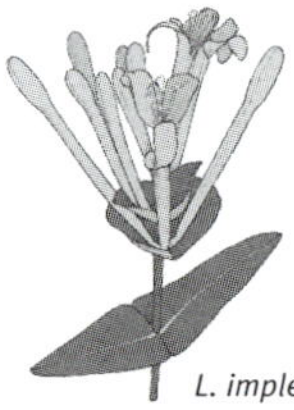

L. implexa

Lonicera implexa **HONEYSUCKLE** An evergreen, much-branched, shrubby and woody climber with bluish, hairless shoots. Leaves entire, to 40 mm long, ovate and pointed, stalkless and cone-shaped on the upper parts of the stems; hairless, dark green and shiny above, and bluish beneath. Inflorescence stemless with up to 9 flowers. Flowers yellowish-white tinged red and fragrant, particularly at night. Fruit a red berry. Common among shrubs and thickets on the *matos* throughout the region.

1

2

3

4

5

6

The following species were previously classified in the family Dipsacaceae, comprising annual or perennial herbs with opposite or whorled leaves; superficially similar to Asteraceae when in flower. Flower-heads spherical or flattened; calyx cup-like; corolla 4-lobed.

Scabiosa atropurpurea **MOURNFUL WIDOW** A medium, hairy, bushy annual or biennial with erect, branched stems. Leaves oblong, untoothed and long-stalked below, pinnately lobed above. Flower-heads lilac, to 30 mm across, becoming oblong in fruit, the outer florets twice the size of the centrals; **involucral bracts not longer than the florets.** Fruits with long calyx teeth. Common in a range of dry habitats in the region, particularly on sandy, maritime waste ground.

S. stellata

Scabiosa stellata A shortly hairy annual with scattered longer hairs. Stems erect to 60 cm, simple or branched. Leaves oblong-elliptical, toothed to pinnately lobed with 4–7 pairs of linear segments, at least above. Flowers borne in distinctive heads made conspicuous by the **cup-like and papery, unlobed (involucral) bracts** in which sit long-pointed, star-like (receptacular) bracts surrounding each floret. Local in disturbed and cultivated or developed areas near the coast.

D. fullonum

Dipsacus fullonum **WILD TEASEL** A familiar, tall biennial to 2 m with robust, prickly, angled stems. Basal leaves forming a large rosette; oblong-elliptical, untoothed and prickly; stem leaves in pairs and fused together on the stem forming a water-collecting cup. Flower-heads pink-purple, large to 80 mm long, and oblong-cylindrical with long, spiny bracts at the base. Local in damp places.

P. rutifolium

Pycnocomon rutifolium **PYCNOCOMON** Similar to *Scabiosa* species when in flower; a robust, hairy or hairless perennial to 2 m with erect, branched stems. Leaves slightly fleshy, once or twice pinnately lobed, reduced above, sometimes unlobed below. Flower-heads white, sometimes flushed pink or yellow, to 20 mm across, the outer florets slightly larger than the centrals; involucral bracts fused together in the lower part forming a cup around the florets. Uncommon, on coastal sands.

The following species were previously classified in the family Valerianaceae, annual or perennial herbs, usually with opposite leaves. Flowers rather small, borne in cymose, often panicle-like heads; calyx inconspicuous; corolla tubular, sometimes with a spur; sometimes 5-lobed; stamens 1–4.

C. calcitrapae

Centranthus calcitrapae A small, hairless annual with simple or branched stems. Leaves rounded to ovate, green flushed with purple, toothed or not, stalked below and stalkless above. Flowers white or pink, small to 2 mm long, tubular and short-spurred, borne in small terminal clusters. Fruit a feathery, persistent calyx. Common on coastal sands, low *matos* and waste places throughout the region.

---

1. *Viburnum tinus*
2. *Lonicera implexa*
3. *Lonicera implexa*
4. *Scabiosa atropurpurea*
5. *Dipsacus fullonum*
6. *Centranthus calcitrapae*

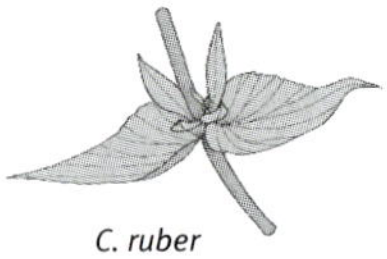
C. ruber

Centranthus ruber **RED VALERIAN** A familiar garden plant; tufted, medium, and somewhat fleshy and waxy. Leaves bluish, ovate, pointed or blunt, clasping the stem above. Flowers usually dark pink-red (sometimes white), tubular and spurred, to 12 mm long, borne in large, slightly fragrant, showy panicles. Fruit single-seeded with a feathery, persistent calyx. Cultivated as an ornamental, and naturalised in the Monchique area.

F. cornucopiae

Fedia cornucopiae **FEDIA** A low to short, much-branched (at an obtuse angle), hairless and slightly fleshy annual. Leaves spatula-shaped to elliptical, stalked and untoothed below; stalkless and toothed above. Flowers purplish-pink, to 15 mm long, borne in stalkless clusters; corolla slender and 5-lobed, slightly 2-lipped; stamens 2–3, with **2 fused together**. Very common in grassy places and fallow land throughout the region.

Valeriana tuberosa A rather fleshy green, purple-flushed perennial with undivided, oblong basal leaves (sometimes shallowly lobed) and pinnately divided stem leaves with deeply dissected linear lobes. Flowers borne in dense, **rounded clusters**; pale pink-white. Fruits have silvery hairs. Generally common on rocky slopes and *matos*.

Order: ASTERALES

CAMPANULACEAE

Annual or perennial herbs with a latex when cut. Leaves alternate. Flowers frequently large and showy, clustered or solitary; corolla bell-shaped, lobed, often blue or purple; stamens 5, fused or free; style solitary. Fruit a capsule.

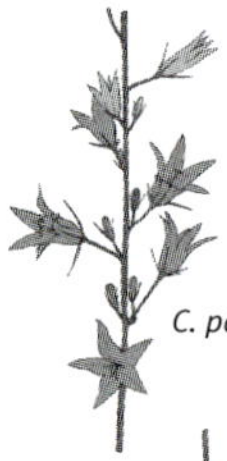
C. patula

C. rapunculus

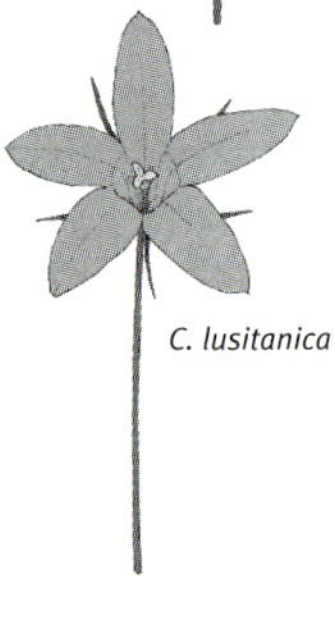
C. lusitanica

Campanula patula **SPREADING BELLFLOWER** A medium, hairless or hairy biennial or perennial with erect or ascending, slender stems. Leaves slightly hairy, ovate below, and toothed above; the upper leaves few and unstalked. Inflorescences lax and branched, carrying flowers that are bell-shaped but open, with lobes as long as their tube, **reddish-violet** to pale blue (rarely white), to 25 mm long; held on erect, slender stems with a **bract in the middle**; buds more or less nodding. Capsules erect with 10 veins. Grassy places on higher ground. ***Campanula rapunculus*** is similar to *C. patula* and easily confused with this species, but has **simple inflorescences** with smaller, always pale blue-mauve flowers, 10–20 mm, held on ascending to erect short stalks that have a **bract at the base**. Roadsides, meadows, cultivated land and other grassy places.

Campanula lusitanica A hairless or hairy annual with long spreading stems to 1 m. Leaves ovate and slightly toothed above; all entire. Flowers bright violet to pale blue, to 20 mm long, usually **borne abundantly in branched inflorescences; flowers borne on long, slender stalks**. Capsules erect. Local on sandy soils.

1. *Centranthus ruber*
2. *Fedia cornucopiae*
3. *Valeriana tuberosa*
4. *Campanula rapunculus*

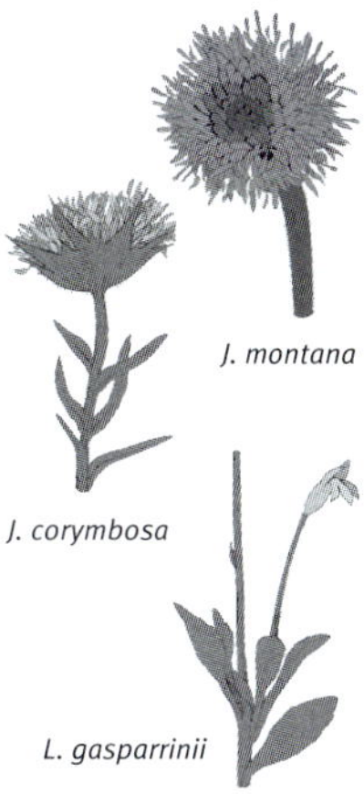

Jasione montana A low, woolly hairy annual or biennial. Stems 5–50 cm, erect or ascending, **leafless beneath the flowers**, simple or branched below. Leaves linear to oblong, undulate and weakly hairy at the margin. Involucral bracts entire, and **shorter than the flowers**. **Flowers borne in dense, blue heads**, small; corolla with 5 narrow lobes; stamens 5; stigmas 2, borne on styles that are hairy in the upper part. Dry, bare and rocky ground; local. ***Jasione corymbosa*** is very similar to *J. montana* but with stems leafy up to the inflorescence, and **involucral bracts as long or longer than the flowers**. Dry, sandy places.

Laurentia gasparrinii A poorly understood species; a small, slender, more or less hairless annual or perennial to 25 cm. Leaves oblong to spatula-like, entire, often forming a basal rosette. Flowers solitary; small to 11 mm borne on stems with 1–2 bracteoles; calyx 5-parted; corolla blue, lilac or white. Distribution improperly known.

1. *Bellis annua*
2. *Bellis perennis*
3. *Bellis sylvestris*
4. *Conyza canadensis*
5. *Evax lusitanica*

ASTERACEAE (COMPOSITAE)

Arguably the largest family of flowering plants (see Orchidaceae) with over 22,000 currently accepted species, spread across more than 1,600 genera and 12 subfamilies, and well-represented in the Mediterranean Basin and in the Algarve region. Herbs or shrubs with alternate, opposite or rosette leaves. Flower-heads typically with an involucre of closely overlapping bracts around the base. Flowers (florets) generally small, borne in congested heads (capitula); those in the centre usually distinct (disc florets) from the those at the edge (ray florets), forming a daisy-like inflorescence. Stamens 5, fused around the style. Ovary inferior; fruit an achene with a feathery appendage (pappus).

Bellis **DAISIES** Small annuals or perennials with basal leaf rosettes, typically daisy-like flowers and achenes without a pappus.

Bellis annua **ANNUAL DAISY** A low, hairless or bristly annual with short, **erect, leafy stems**. Leaves spatula-shaped, toothed or not, stalked below. Flower-heads white with a yellow disc, to 15 mm across, the rays sometimes tinged purple. Common in damp grassy as well as dry open habitats throughout the region. ***Bellis perennis*** **(DAISY)** is a similar perennial with a **stemless**, dense basal rosette, **leaves abruptly tapered at the base** into a petiole and larger flowering heads to 30 mm across. Common on lawns, roadsides and other grassy habitats. ***Bellis sylvestris*** is similar to *B. perennis* but with **3-veined leaves gradually tapered at the base** into a petiole. **Flower-heads large**, to 40 mm across, borne on long, stout stalks to 30 cm; ray florets white but sometimes tinged purple on both sides. Fruit compressed and bristly. Woods, thickets and shady roadsides; local. Occurs on tracks around Picota.

Conyza **FLEABANES** Herbs with alternate, simple leaves dense along erect stems. Flowers inconspicuous; most florets female; cosexual florets few and yellow. Achenes with a pappus of hairs.

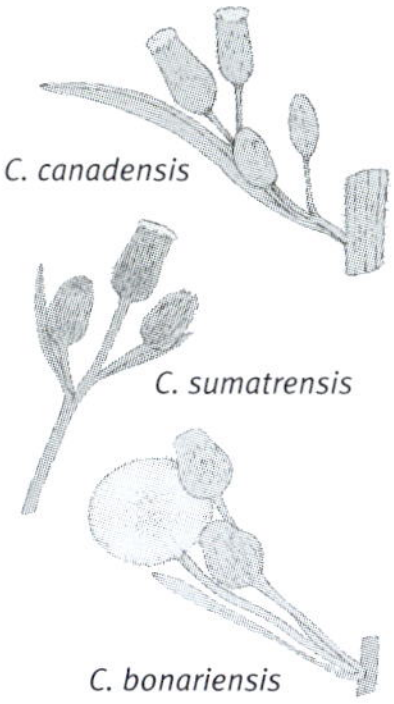

Conyza canadensis **CANADIAN FLEABANE** A tall, hairy annual to 1.5 m. Leaves alternate, narrowly oblong, stalked and yellow-green, often withered below and deciduous before flowering. Flower-heads very small, 2.5–5 mm, whitish, borne abundantly in lax, branched clusters terminally. **Flower bracts hairless** or nearly so, and brownish within. A weed native of North America, naturalised on waste ground in towns. ***Conyza bonariensis*** is native to tropical America and is similar but taller (to 2.5 m) and more densely hairy, with larger flower-heads (to 10 mm) borne in an inflorescence in which longer branches overtop the main axis; **bracts densely hairy**. ***Conyza sumatrensis*** is native to the Americas and is similar to *C. bonariensis*, with hairy **bracts that lack long hairs at the apex** and that are reddish within.

Evax Small annuals with alternate leaves. Flower-heads in clusters without ray florets, forming a domed, cushion-like inflorescence; involucral bracts numerous.

Evax pygmaea A very small, grey-felted annual to 40 mm high, branched at the base. Leaves oblong, **narrow** and blunt, to 15 mm long and 5 mm wide, and forming a rosette; upper leaves 2–3-times longer than flower-heads; all leaves without distinct stalks. Flowers borne in very compact clusters to 35 mm across; brownish-yellow. Fairly common on dry, bare, and stony places, as well as in low coastal *matos*. ***Evax lusitanica*** is very similar to *E. pygmaea*, but **virtually stemless,** with **broader rosette leaves** to 8 mm wide; greyish-green and rounded at the tip. Southeast of the region; local on cliff tops and pathways, varying in abundance with seasonal rain.

Filago **CUDWEEDS** Downy and woolly annuals with alternate, untoothed leaves, and inconspicuous flowers borne in small, lateral or terminal heads; achene with a pappus. A taxonomically difficult genus.

F. vulgaris

Filago vulgaris **COMMON CUDWEED** An erect, white-woolly annual to 30 cm, branched or unbranched below, always branched above in 2–3 forks. Leaves erect, lanceolate, to 20 mm long, wavy-edged and untoothed. **Flower-heads borne as dense rounded clusters** of up to 30 flowers in the branch axils and terminally, **not exceeded by the leaves** immediately below; inner flower bracts with a transparent bristly tip. Common on bare disturbed, sandy ground. ***Filago lutescens*** is similar to *F. vulgaris* but with erect, **yellow-woolly stems** and broader, not wavy, bristle-tipped leaves. Flower-heads with up to 20 flowers, over-topped by some leaves; flower bracts yellowish, with prominent, erect, **red-brown or purplish bristle-tips**. Local in sandy habitats.

F. pyramidata

Filago pyramidata **BROAD-LEAVED CUDWEED** Rather similar to *F. vulgaris*; a greyish-white annual always branched and **spreading from the base**, often almost prostrate. Leaves ovate and bristle-tipped. Flower-heads borne as dense clusters of up to 20 flowers in the branch axils and terminally, generally **exceeded by the leaves** immediately below; **outer flower bracts with yellow points**, curving inwards in fruit. Open fallow and sandy waste places, and sandy cliff tops; common.

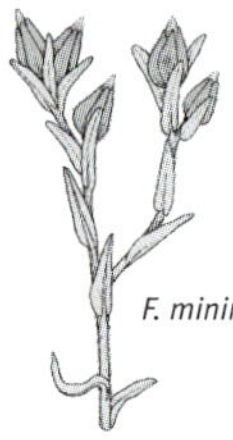

F. minima

Filago minima **SMALL CUDWEED** A low to short, grey silkily haired annual to less than 30 cm tall. Stems very slender and branched above the middle. Leaves linear-lanceolate, to 10 mm long. Flower-heads ovoid to pyramidal, small to 3.5 mm long, **not over-topped by the leaves** immediately below; **outer flower bracts woolly at the base**, but yellow and **hairless at the tip**. Open fallow and sandy waste places; local. ***Filago gallica*** is similar to *F. minima* but with linear, **thread-like leaves to 25 mm long**, and **flower-heads over-topped by the leaves**; flower-bracts woolly and yellowish at the tip. Local, in habitats similar to those of *F. minima*.

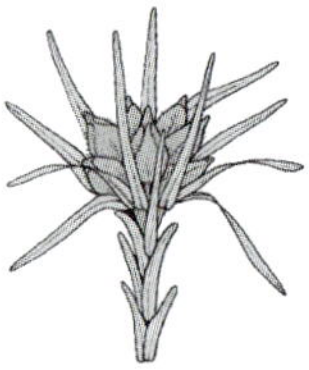

F. gallica

***Gnaphalium sylvaticum* (= *Omalotheca sylvatica*)** **HEATH CUDWEED** A grey-hairy perennial to 50 cm tall with non-flowering shoots and densely leafy stems. Leaves lanceolate to linear, decreasing in size along the stem, woolly beneath. Inflorescence lax, and interrupted below; capitula yellowish and small; involucre compact and narrowly bell-shaped; involucral bracts dark brown. Heaths and scrub; local.

G. luteo-album

Gnaphalium luteo-album A short annual rather like *Filago lutescens* but **white-woolly**, unbranched or with branches spreading from the base then soon erect, to 30 cm. Leaves broadly lanceolate, blunt, and running down the stem. Flower-heads terminal and densely ovoid, and **not overtopped by their surrounding leaves** (at least when mature); bracts elliptical, **shining, straw-yellow**, only the outermost woolly below, and not bristle-tipped. Florets yellowish, with red stigmas.

---

1. *Filago pyramidata*
2. *Gnaphalium luteo-album*
3. *Helichrysum stoechas*
4. *Helichrysum stoechas* in bud
5. *Helichrysum italicum*
6. *Helichrysum italicum*

Helichrysum **CURRY PLANTS** Dwarf greyish aromatic shrubs with alternate, untoothed leaves. Flowers borne in dense clusters with papery bracts. Achene with a pappus.

H. stoechas

Helichrysum stoechas A spreading, much-branched dwarf shrub, to 50 cm tall. Leaves white-felted, linear, untoothed and slightly aromatic (curry-scented) when crushed, to 30 mm long, with down-turned margins. Flower-heads yellow, borne in clusters to 30 mm across; the series of bracts (involucre) bright shiny yellow and **spherical** in flower; the outer flower bracts papery white and hairless or hairy at the base. Very common on fixed sand dunes and low coastal *matos* throughout the region.

H. italicum

Helichrysum italicum Similar to *H. stoechas*; a spreading, much-branched dwarf shrub to 50 cm tall. Leaves white-felted when young but later hairless, linear, untoothed and strongly aromatic (curry-scented) when crushed, to 30 mm long, with down-turned margins. Flower-heads yellow, borne in clusters to 80 mm across; the series of bracts (involucre) **dull mustard-yellow** and **oblong** to narrowly bell-shaped in flower; the inner bracts at least 5-times longer than the outer. Common on fixed sand dunes and low coastal *matos*.

1

2

3

4

5

6

Phagnalon Grey dwarf shrubs with alternate leaves and flower-heads solitary at the tips of the branches; flower-heads with densely overlapping bracts. Achene with a pappus.

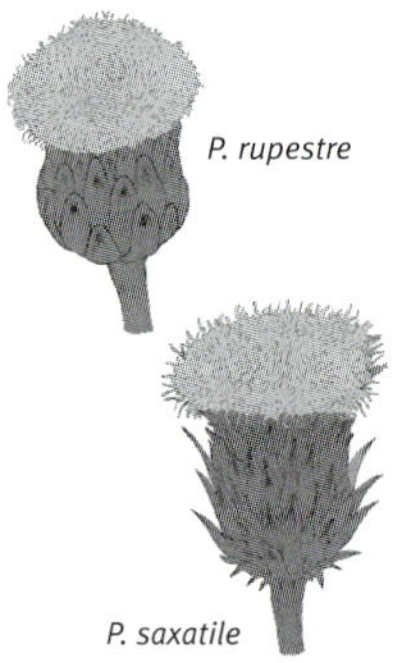

Phagnalon rupestre A small shrub to 30 cm tall with erect to ascending, white-felted stems. Leaves narrowly ovate and more or less toothed, small (10–40 mm), green or whitish above and white-felted below, with down-turned margins. Flower-heads solitary, distinctly **long-stalked**, yellowish, oblong, to 15 mm; flower bracts brownish, membranous, somewhat hairy and closely overlapping. Fairly common on cliffs, sun-baked rocky ground and low coastal *matos*. **Phagnalon saxatile** is similar but taller (to 60 cm). **Leaves linear**, sometimes broadest above the middle, to 60 mm, **green** (not normally white-felted) above and white-felted below. Flower bracts with somewhat wavy margins, the tips often slightly recurved. Local in rocky habitats.

Dittrichia Annual or perennial herbs or small shrubs with simple, alternate leaves. Flower-heads small to medium. Achenes abruptly contracted below the pappus.

D. viscosa

Dittrichia viscosa A densely glandular, sticky perennial to 1 m with stems woody at the base. Lower leaves bright green, linear, scarcely and laxly toothed; upper leaves stalkless and semi-clasping the stem. Flower-heads daisy-like and bright yellow, to 20 mm with rays to 8 mm long, much exceeding the flower bracts. Common or abundant in all disturbed and damp habitats in lowland areas throughout the region.

Pulicaria Annual or perennial herbs with simple alternate leaves. Flower-heads small to large, yellow and daisy-like. Achenes with scales around the pappus.

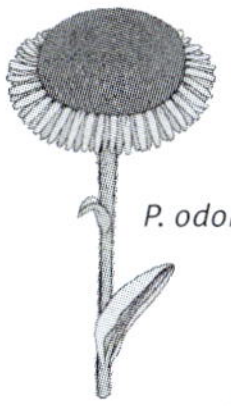
P. odora

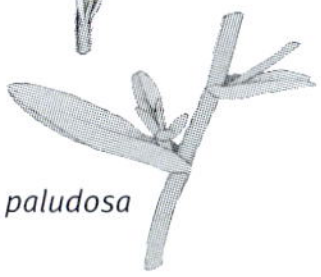
P. paludosa

Pulicaria odora A perennial with a short stock and stems to 70 cm, unbranched or sparingly so. Basal leaves ovate-lanceolate, short-stalked; upper leaves stalkless and semi-clasping the stem. Capitula few in number, to 30 mm across, hemispherical, with long, bright yellow ray florets, borne on stalks 30 mm to 15 cm long, thickened above with several bracts; involucral bracts linear and long. Achenes hairy, the pappus of scales surrounding 10–12 hairs. Locally frequent in seasonally damp places. ***Pulicaria paludosa*** is similar but with narrow, rigid, **linear leaves**, smaller capitula, and achenes with erect to spreading hairs. Habitats similar to those of *P. odora*.

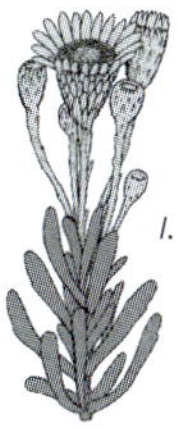
I. crithmoides

Inula crithmoides A sprawling perennial with a woody base and **fleshy**, crowded linear leaves that are untoothed or 3-toothed at the apex. Flower-heads yellow with a golden disc, borne in flat-topped clusters. Coastal habitats.

Palensis spinosa A rather robust, softly hairy annual with rigid stems, woody at the base with branches overtopping the main stem. Leaves elliptical, blunt-tipped and stalked below, unstalked and semi-clasping the stem above. Flower-heads daisy-like and bright yellow with a large disc to 20 mm across; **flower bracts spine-tipped and twice as long as the ray florets**; inner flower bracts papery and not spine-tipped. Fairly frequent on coastal *matos*, dry roadsides and rocky places.

A. maritimus

Asteriscus maritimus A short, rather compact perennial or subshrub with mat-forming stems that are woody at the base and much-branched; **not aromatic**. Leaves spatula-shaped, short-stalked and untoothed. Flower-heads numerous, bright yellow and daisy-like, to 40 mm across, surrounded by leafy flower bracts not that are spine-tipped. Common to abundant on coastal cliff-tops and rocks; rare inland in the region. ***Asteriscus aquaticus*** is similar to *A. maritimus* but an **aromatic annual**, with erect stems, and **flower bracts greatly exceeding the ray florets**. Damp and sandy places, waste places and roadsides.

---

1. *Phagnalon rupestre*
2. *Phagnalon saxatile*
3. *Dittrichia viscosa*
4. *Palensis spinosa*
5. *Asteriscus maritimus*

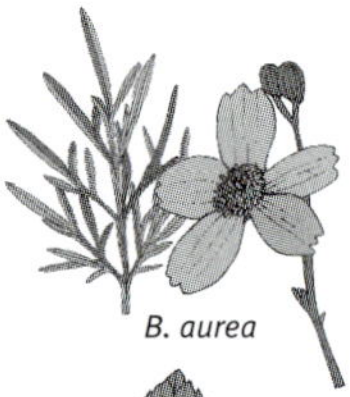

B. aurea

Bidens aurea BIDENS A virtually hairless, medium to tall, slender perennial to 1 m. Leaves linear-lanceolate or twice pinnately divided with linear lobes; toothed. Flower-heads held erect, daisy-like, yellow, white or mottled both, to 60 mm across, with only **5 or 6 large, broad grooved ray florets**; flower bracts all similar and much shorter than the ray florets. Native to Central America, naturalised in damp places.

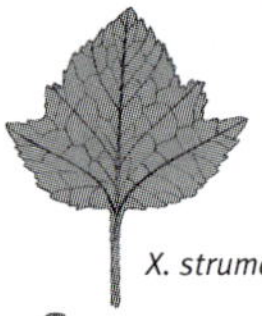

X. strumatrium

Xanthium strumatrium A stiffly branched annual, to 1.2 m, without spines. Leaves alternate and heart-shaped at the base, unlobed or with 3–5 lobes. Flower-heads greenish and with male and female flowers borne separately in **lateral clusters**. Naturalised in damp habitats.

S. rosmarinifolia

Santolina rosmarinifolia A dense, aromatic, grey dwarf shrub to 70 cm tall. Leaves small, grey, narrowly oblong and lobed; lobes few and distant. Flower-heads yellow, borne terminally on numerous leafless stalks thickened towards the top. Dry, rocky habitats.

Anthemis Slightly hairy, often aromatic herbs or dwarf shrubs with alternate, divided (sometimes feathery) leaves.

A. maritima

Anthemis maritima A dwarf hairless or slightly hairy subshrub to 50 cm tall with stout stems that root at the base; leafy above. Leaves pinnately lobed, **fleshy** and gland-dotted. Flower-heads daisy-like with white rays and a yellow disc, borne on **long, slender stalks** to 10 cm. Fairly common on coastal sands.

A. arvensis

A. tinctoria

Anthemis arvensis CORN CHAMOMILE A short to medium, hairy, aromatic herb with spreading or ascending branched, downy stems to 50 cm. Leaves to 50 mm long, ovate in outline, once to 3-times pinnately lobed with narrow, pointed segments; **woolly below**, especially when young. Flower-heads daisy-like with white rays and a yellow disc, to 40 mm across; flower bracts with brown, papery margins. Disturbed, cultivated and fallow land; frequent. ***Anthemis tinctoria*** is similar to *A. arvensis* but with leaves that have broader segments and are less divided. **Flowers all yellow**, produced in large numbers in flat-topped clusters. Habitats similar to those of *A. arvensis*.

Achillea Perennial herbs with pinnately divided (sometimes feathery) or undivided, alternate leaves. Flower-heads congested into umbel-like flat-topped clusters.

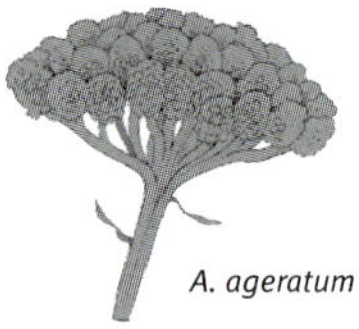

A. ageratum

Achillea ageratum An erect, simple or branched, hairy perennial with stems woody at the base. Leaves to 40 mm, lanceolate to linear in outline but pinnately divided; those of the non-flowering stems divided only in the upper part, **linear-toothed and entire below**. Flower-heads yellow. Local in damp habitats.

C. mixtum

Chamaemelum mixtum A short to medium, hairy, chamomile-like annual to 60 cm, normally with much-branched stems. Leaves oval in outline, once to twice pinnately divided below, deeply toothed or **pinnate and unstalked above**; leaf lobes linear-lanceolate and toothed or not. Flower-heads daisy-like with spreading white rays and a yellow disc; ray florets 3-toothed at the tip; flower bracts with a wide, pale brown membranous margin. Cultivated, fallow and sandy waste ground, and coastal sands.

Chamaemelum fuscatum is similar to *C. mixtum* but shorter (to 30 cm), **hairless**, less densely branched, with **flower-heads nodding** in bud; **flower bracts with blackish-brown margins** and tips. ***Chamaemelum nobile*** is a creeping, **strongly aromatic** perennial with daisy-like flower-heads. Sandy habitats; local.

Anacyclus Chamomile-like annual or perennial herbs with alternate leaves that are once or twice pinnately divided. Outer achenes 2-winged, inner achenes unwinged; pappus absent.

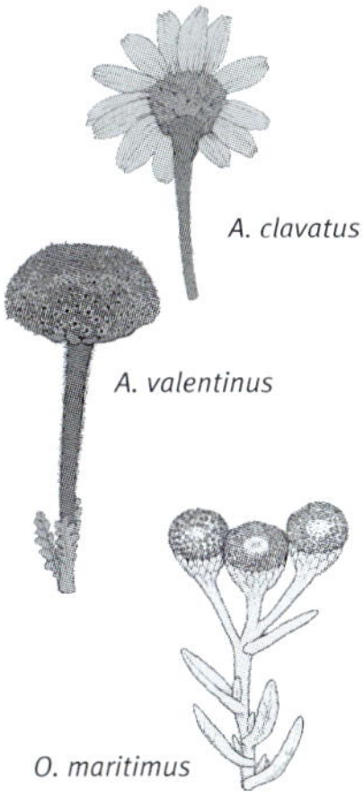

A. clavatus

A. valentinus

O. maritimus

Anacyclus clavatus A short, widely branched, hairy annual. Leaves alternate, twice to 3-times pinnately lobed with linear segments. Flower-heads daisy-like with **white** recurved rays and a yellow disc; flowers solitary on **stalks distinctly swollen below the fruiting head**; flower bracts with a narrow whitish or purplish margin. Sandy ground and coastal rocks. ***Anacyclus radiatus*** is similar to *A. clavatus* but taller (to 60 cm), with **flowers all yellow**, the rays purplish beneath. Sandy places. ***Anacyclus valentinus*** is similar to *A. clavatus* but with **flower-heads with very short rays**, that do not exceed the involucre (appearing rayless). Habitats similar to those of *A. clavatus*.

Otanthus maritimus COTTONWEED A short, densely **white-woolly**, spreading perennial with robust, ascending stems. Leaves oblong to lanceolate, untoothed or blunt-toothed, fleshy and unstalked. Flower-heads small to 10 mm across, yellow, rayless and button-like, borne in lax, flat-topped clusters; flower bracts white-woolly. Fixed dunes and coastal sands; local.

Chrysanthemum Annuals with solitary, or 2–5, capitula on the branches; receptacle without scales. Achene without a pappus.

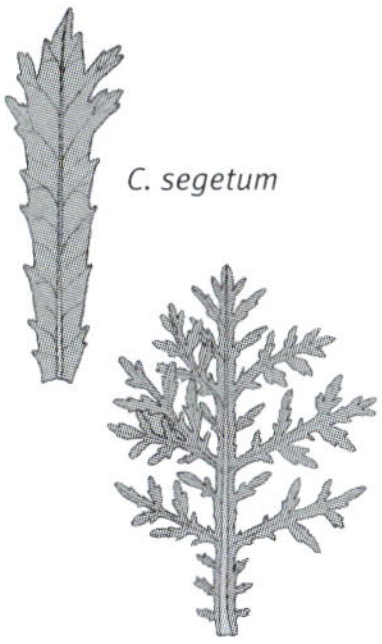

C. segetum

C. coronarium

Chrysanthemum segetum CORN MARIGOLD A short to tall, green, hairless, slightly fleshy annual with erect to ascending, branched or unbranched stems. Leaves alternate, narrowly ovate in outline, deeply toothed, at least below. Flower-heads bright yellow and daisy-like, rather large to 55 mm across with a flat disc. Common on disturbed fallow and cultivated land, roadsides and coastal waste places.

Chrysanthemum coronarium CROWN DAISY Rather similar to *C. segetum*; a slightly hairy annual with **leaves twice pinnately lobed**. Flower-heads yellow and daisy-like, large from 30–60 mm across, with **rays white in the upper half**. Very common to abundant on disturbed land, roadsides and coastal waste places throughout the region.

1. *Anthemis maritima*
2. *Chrysanthemum segetum*
3. *Chrysanthemum coronarium*

C. myconis

L. repandum

Coleostephus myconis Similar to *Chrysanthemum*; a short to medium, sparingly branched, virtually hairless annual. Leaves ovate, **regularly toothed (not lobed)**; lower leaves tapered gradually into a stalk, upper leaves unstalked and semi-clasping the stem. Flower-heads yellow and daisy-like, to 22 mm across with yellow or paler rays and yellow **disc florets**. Cultivated, fallow and waste ground. ***Coleostephus clausonis*** is similar but with **prostrate stems** and **leaves irregularly toothed**; Habitats similar to those of *C. myconis*.

Lepidophorum repandum A *Chrysanthemum*-like hairless annual with sparingly branched stems, to 50 cm. Leaves spatula-shaped to oblong, shallowly toothed (**not lobed**), long-stalked below, stalkless above. Flower-head daisy-like and yellow, to 40 mm across. Fallow land; distribution improperly known in the region.

Daveaua anthemoides A short, hairless annual to 30 cm. Leaves alternate, pinnately divided with **very narrow, thread-like linear segments**. Flower-head solitary, to 35 mm across. Waste places and cultivated ground; distribution improperly known in the region.

Leucanthemum vulgare **OX-EYE DAISY** A short to tall, variably hairy, clump-forming perennial with leafy stolons and erect, ridged stems. Leaves alternate, oblong, toothed and stalked below, unstalked and clasping the stem above. Flower-heads large and daisy-like, to 50 mm across with long, white rays and a yellow disc. Locally frequent in undisturbed grassy places and meadows.

1. *Leucanthemum vulgare*
2. *Cotula coronopifolia*
3. *Senecio gallicus*
4. *Senecio lividus*

Aster tripolium **SEA ASTER** A short to medium, hairless annual with reddish, erect or ascending stems, branched from the base. **Leaves fleshy**, linear to linear-lanceolate, rounded in cross-section and clasping the stem; unstalked above. Flower-heads with bright blue-purple rays and a yellow discs, to 20 mm across, borne in large, flat-topped inflorescences. Local in maritime environments.

Cotula Small annual or perennial herbs with entire or pinnate, alternate leaves. Flower-heads borne in leaf axils or terminally, without rays. Achene without a pappus.

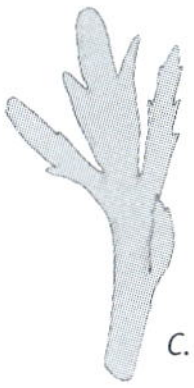

C. coronopifolia

Cotula coronopifolia A small, hairless, often patch-forming, annual to 30 cm. Leaves rather fleshy, linear and entire or scarcely toothed, and clasping the stem at the base. Flower-heads all yellow, to 10 mm in diameter, without rays; long-stalked and often nodding. Achenes strongly compressed and winged (outer disc florets) or unwinged (inner disc florets). Native to South Africa; locally naturalised in damp and seasonally flooded habitats.

Artemisia Herbs or low shrubs, often highly aromatic. Leaves alternate. Flower-heads small and often nodding in a lax inflorescence. Achenes without a pappus (or virtually so). A large genus that is poorly represented in the region.

A. campestris

Artemisia campestris A scarcely aromatic, low, mat-forming shrub with a stout, woody stock and numerous non-flowering shoots. Stems ascending and persistently, **silkily hairy**. Leaves silky when young, later virtually hairless; twice to 3-times pinnately divided below, simple and stalkless above. Flower-heads pale green, ovoid, short-stalked and erect or spreading; involucre hairless or virtually so; bracts with a wide, papery margin; corolla yellowish or reddish. **Subsp. *maritima*** is the form found in the region; it has less hairy stems and leaves, **fleshy leaf lobes**, and broad panicles of flowers. Local but in large numbers where it occurs, only on maritime sands.

Doronicum Herbaceous perennials with tubers or stolons. Leaves simple and alternate. Flower-heads with a hairy receptacle.

Doronicum plantagineum A simple or sparingly branched, slender annual to 80 cm; hairless below, hairy above. Basal leaves pale green and ovate-elliptical, to 60 mm long, untoothed or weakly so; upper leaves ovate to triangular and clasping the stem. Flower-heads daisy-like with an orange-yellow disc; long, slender, yellow rays, to 50 mm across, solitary on glandular-hairy stalks; involucral bracts linear. Dry rocky places, possibly only in Rocha da Pena.

Senecio A large and very diverse genus of herbs and shrubs with alternate leaves. Flower-heads often numerous, borne in flat-topped clusters. Achene with a white or greyish pappus.

S. gallicus

S. lividus

Senecio gallicus A short, hairless or slightly hairy annual usually branched from the base. Leaves fleshy, pinnately divided with lobed segments; stalked below, unstalked above and **clasping the stem**. Flower-heads to 22 mm across, bright yellow, borne in lax clusters; rays to 8 mm long. Common in the region on coastal sands. ***Senecio lividus*** is similar to *S. gallicus* but with taller, normally **unbranched stems** and sparsely glandular-hairy, with **smaller flower-heads** (to 10 mm across) with black-tipped bracts and short, **recurved rays**. Similar habitats and sometimes co-occurring with *S. gallicus*, though generally less common.

Senecio vulgaris GROUNDSEL A short, more or less hairy, rather succulent annual to 30 cm (usually less) with weak, sparingly and irregularly branched stems. Leaves coarsely lobed and toothed, ovate in outline and short-stalked below, and semi-clasping the stem at the base. **Flower-heads small and without (or with few) rays**, to 5 mm across, yellow; involucre cylindrical with black-tipped bracts. A common weed in a range of disturbed natural and urban habitats.

S. viscosus

Senecio viscosus STICKY GROUNDSEL A small, **very sticky**, rather unpleasant-smelling annual to 60 cm (usually less). Stems weak, freely branched or sparingly branched. Leaves dark grey-green, densely glandular-hairy and sticky, deeply and regularly lobed, short-stalked below and unstalked (but not clasping) above. Flower-heads to 12 mm, with 13 pale yellow, recurved ray florets; borne in a large, irregular corymb. Local on wasteland and in open stony and sandy places.

Senecio jacobaea (= Jacobaea vulgaris) RAGWORT A large biennial or weak perennial to 1.5 m tall. Stems reddish, robust and erect, more or less hairless or slightly woolly (floccose) and unbranched below; **furrowed**. Basal leaves large (to 20 cm long), **deeply pinnately lobed with a large blunt end-lobe** and stalked, often withered at flowering time; upper leaves semi-clasping the stem; all leaves slightly hairy below. Flower-heads to 20 mm across, bright yellow and numerous, borne in **flat-topped clusters**. Involucre more or less hairless. Rather local, in grassy habitats.

Senecio angulatus A creeping, **succulent** perennial with stems to 2 m long and bluntly lobed, shiny green leaves. Flowers yellow, typically *Senecio*-like and scented. A native of South Africa; commonly planted in the region.

Calendula Annual or perennial, often aromatic herbs with alternate, undivided leaves. Flower-heads daisy-like with yellow or orange ray florets; bracts in 1–2 rows. Achenes strongly curved, with a pappus.

C. suffruticosa

C. arvensis

C. officinalis

Calendula suffruticosa A short to medium perennial, sometimes with a woody stock. Leaves lanceolate, somewhat fleshy, pointed, with few, distant teeth; glandular. Flower-heads large, to 40 mm across, bright yellow, the rays sometimes red-tipped, to 20 mm long. Fruiting head with a **conspicuous outer row of spreading (weakly curved) achenes**. Fairly common in rocky and sandy coastal habitats, especially sea cliffs. **Subsp. *tomentosa*** is **white-hairy**, has erect stems and has leaves that are not fleshy. **Subsp. *algarbiensis*** has **smaller flowers** (20–35 mm across), and minutely but **densely glandular-hairy leaves**. **Subsp. *lusitanica*** is **scarcely glandular-hairy**.

Calendula arvensis FIELD MARIGOLD A short, slender, thinly hairy, ascending or spreading annual. Leaves oblong and finely toothed. Flower-heads daisy-like and **small** (to 27 mm across), yellow-orange throughout or with a brownish disc. Fruiting head with an outer row of beaked and strongly incurved achenes. Common on fallow ground and open *matos*. ***Calendula officinalis*** is similar but with **larger flowers** (to 70 mm across), yellow-orange with an orange or brownish disc. Widely cultivated as an ornamental, sometimes naturalised.

A. calendula

Arctotheca calendula An annual with spreading, leafy stems to 40 cm long. Leaves 7–20 cm long, divided, roughly hairy above and white softly hairy beneath. Flower-heads to 50 mm across; outer involucral bracts with a membranous margin and often with a terminal appendage; inner bracts membranous; rays long, distally deep yellow and pale yellow at the tip; disc florets blackish green. Common in bare places and sandy waste ground.

1. *Senecio vulgaris*
2. *Senecio jacobaea*
3. *Calendula suffruticosa*
4. *Calendula arvensis*
5. *Calendula arvensis* in fruit
6. *Arctotheca calendula*

Gazania rigens A bushy perennial to 40 cm high with spreading stems, woody at the base. Leaves dark green, oblong, entire, and tapered gradually at the base into a long stalk, equalling the lamina; densely white-hairy below. Flowers large and showy, to 80 mm across; rays bright orange or white with a basal black patch; disc orange. Cultivated as an ornamental and occasionally naturalised.

Carlina Spiny, thistle-like annuals with lobed leaves. Flower-heads with tubular florets but without ray florets; bracts conspicuous and spreading, resembling rays superficially.

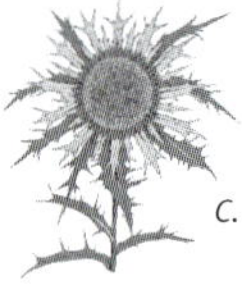
C. corymbosa

C. racemosa

Carlina corymbosa FLAT-TOPPED CARLINE THISTLE A pale, whitish or bluish short to tall, erect, spiny perennial. Main stems usually solitary or few. Leaves linear-lanceolate, toothed and pinnately lobed, wavy and spiny-margined, clasping the stem. **Flower-heads stalked**, borne terminally in flat-topped clusters, to 20 mm across, with yellow florets and golden brown, long, narrow, spiny bracts. Common and widespread across the region on dry stony slopes, sea cliffs and *matos*. **Carlina racemosa** is similar but with larger, unstalked flower-heads to 15 mm across that are **distinctly overtopped** by 1–2 surrounding flowering branches, which arise from immediately below. Grassy and stony open habitats and fixed dunes; locally common.

1. *Carlina corymbosa*
2. *Carlina racemosa* in bud
3. *Atractylis cancellata*
4. *Carduus tenuiflorus*
5. *Galactites tomentosa*
6. *Galactites tomentosa*

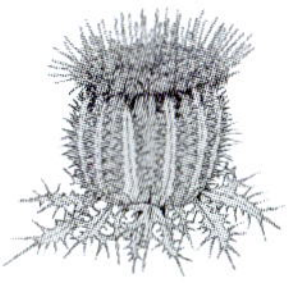

A. gummifera

Atractylis gummifera A thistle-like, low, robust, **stemless** perennial with leaves in a lax rosette. Leaves oblong in outline but deeply divided into narrow, spiny segments; stalks clasping at the base. Flower-heads large and pink-purple without ray florets, with an involucre to 70 mm in which the outer bracts bear 3 apical spines; inner bracts with a single brown spine. Local in dry fields, roadsides and fallow land.

A. cancellata

Atractylis cancellata A slender, thistle-like perennial with leaves in a lax rosette and with **stems 3–30 cm**. Leaves hairy, oblong in outline, toothed and shortly spine. Flower-heads large and pink-purple without ray florets, with an involucre to 20 mm, surrounded by **upper leaves forming a cage-like structure**; middle bracts with all spines similar. Local in dry fields, roadsides and fallow land.

Echinops strigosus GLOBE THISTLE A spiny, bushy and rigid perennial to 50 cm. Leaves elliptical in outline but deeply divided with linear segments; dense, stiff hairs above, softly white-hairy beneath. **Flower-heads spherical**, bluish, to 70 mm in diameter. Local in dry, scrubby places.

Carduus Annual to biennial thistles with spiny-winged stems, and alternate, spine-toothed leaves. Flower-heads rounded or cylindrical, often brush-shaped; ray florets absent.

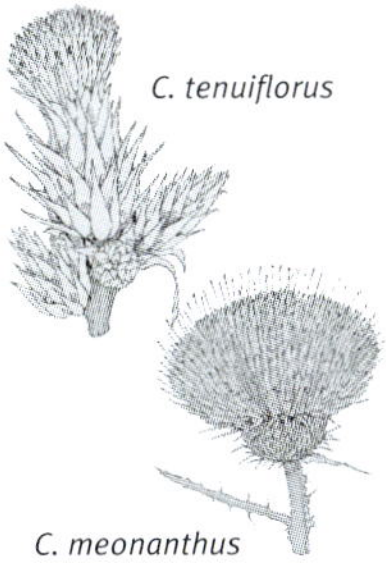

C. tenuiflorus

C. meonanthus

Carduus tenuiflorus SLENDER THISTLE An erect, narrowly branched biennial to 1 m tall with **stems broadly winged** up to the flower-heads; very cottony and greyish. Flower-heads slender, to 20 mm long and 5 mm wide, borne in dense terminal clusters; florets pale pink-red; bracts at least 1.5 mm wide. Fairly common on higher ground in dry open woods, and bare and disturbed habitats. ***Carduus meonanthus*** is a maritime biennial similar to *C. tenuiflorus* but with **virtually hairless stems** and narrower bracts. ***Carduus bourgeanus*** is an annual to 40 cm distinguished by its wide, **bell-shaped flower-heads** at least 10 mm wide.

C. bourgeanus

Cirsium vulgare SPEAR THISTLE Rather similar to *Carduus* species; an erect biennial to 1.5 m with cottony stems that have interrupted spiny wings. Basal leaves to 30 cm long, deeply pinnately lobed with segments forked and spiny, and a **single long, pointed end-lobe**; upper leaves smaller; all leaves prickly. Flower-heads ovoid, to 50 mm long, with cottony bracts; the outer bracts with long spine-tips; florets reddish-purple. Disturbed habitats; local.

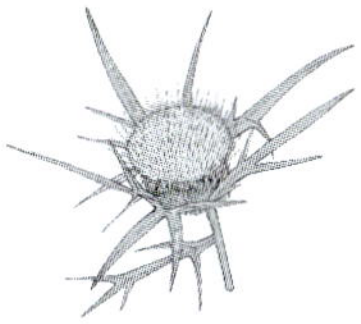

N. syriaca

Notobasis syriaca SYRIAN THISTLE A medium, thistle-like annual with rigid **stems not spiny-winged**. Leaves alternate, dark green with paler veins; narrowly elliptical, pinnately lobed with spine-tipped, narrow triangular lobes, reduced and **clustered** around the stalkless flower-heads above. Flower-heads to 23 mm, purple; bracts with spine-tips. Field margins, fallow land and roadsides.

G. tomentosa

Galactites tomentosa GALACTITES A stiffly erect perennial to 1 m with conspicuously **white-veined and variegated dark green leaves**; alternate, oblong, pinnately lobed with spiny lobes, white-downy beneath. Flower-heads pale purple, to 20 mm, solitary or in clusters, the **outer ray florets long, even and spreading**; flower bracts tapered abruptly into spine tips, white-downy. Very common in a range of dry and disturbed habitats in the region.

Onopordum Stout perennial thistles with spiny-winged stems, often with cobweb-like hairs. Leaves spiny-margined. Flower-heads large and purple or white; all florets tubular and deeply 5-lobed.

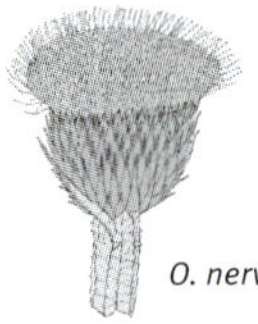

O. nervosum

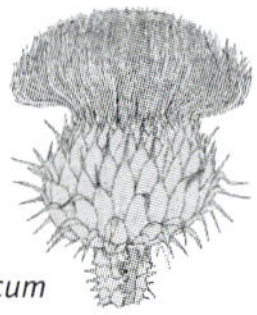

O. illyricum

Onopordum nervosum A **large**, thistle-like perennial to 2.7 m. Stems yellowish hairy, **with broad, veined wings to 20 mm** and long spines to 10 mm; branched at acute angles. Leaves large, to 50 cm, oblong-lanceolate and stalkless, green with paler veins, more or less hairless above and sparsely hairy and strongly veined below; divided with 6–8 pairs of lobes, each with a long spine-tip. Flower-heads pink, to 50 mm, with florets to 30 mm; bracts spine-tipped. Local. ***Onopordum illyricum*** is similar but smaller, to 1.3 m, with **narrower wings not veined**, remotely lobed leaves, and larger, purplish flower-heads to 60 mm across, with florets to 35 mm. ***Onopordum macracanthum*** is similar to both species but **cottony-white** throughout, with **sparsely spiny stems**.

Cynara Stout perennial thistles with leaves in a basal rosette or alternate; deeply divided with spiny segments. Flower-heads solitary or sparingly; purplish, blue or white.

C. humilis

Cynara humilis A large, bushy, perennial thistle to 80 cm with numerous basal leaves; stems white-hairy and unwinged. Leaves lanceolate in outline, **deeply twice pinnately divided into linear segments with down-turned margins**; hairless and somewhat shiny above, white-hairy beneath; lower leaves short-stalked and upper leaves stalkless. Flower-heads very large; involucre to 60 mm long, brownish when mature with purplish **bracts with prominent spine-tips**; florets purplish. Common on dry fallow ground on heavy soils.

C. cardunculus

Cynara cardunculus Rather similar to *C. humilis*; a large, bushy, perennial thistle to 1 m with numerous basal leaves. Stems white-hairy and unwinged. Leaves thick, lanceolate in outline, deeply once to twice pinnately lobed into **lanceolate** (not narrow and linear), **flat segments**; toothed, shortly and sparsely hairy and green above, white-hairy beneath; lower leaves short-stalked and upper leaves stalkless; **long spines to 35 mm clustered at the base of each leaf segment**. Flower-heads large; involucre to 60 mm long, with bracts narrowed into spreading spine-tips; florets violet-purple. Common on dry stony, fallow and waste places on heavy soils. ***Cynara algarbiensis*** is similar to *C. cardunculus* but with leaves **blue-white** and rather hairy above (not bright green), white-hairy and indistinctly veined beneath, with **variably short marginal spines** to 6 mm. Flower-heads blue-purple. Fairly common on disturbed and fallow ground or roadsides on clay soils.

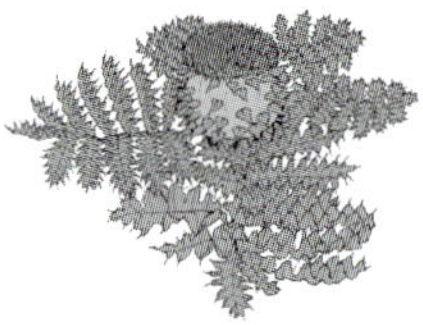

C. tournefortii

Cynara tournefortii A **low, stemless**, robust, spiny, thistle. Leaves to 20 cm, oblong-lanceolate, virtually hairless above, grey-hairy below, divided into lobes that are toothed at the base (with yellow spine-tips), untoothed towards the tip. **Flower-heads large and solitary in the centre of the rosette**; involucre to 80 mm; florets blue. Waste and fallow ground on clay soils.

Silybum marianum **MILK THISTLE** A robust, medium to tall, weakly spiny biennial to 1.5 m. Basal leaves oblong, pinnately lobed and prominently **white-veined and variegated**, virtually hairless and stalked beneath; stem leaves smaller, less lobed and clasping the stem. Flower-heads purple, to 40 mm, solitary and terminal; flower bracts terminating in long, stout spines to 7 mm. Very common and widespread on fallow, cultivated, waste ground and field margins.

1. *Cynara humilis*
2. *Cynara cardunculus*
3. *Cynara algarbiensis*
4. *Silybum marianum*

Cheirolophus Very similar to *Centaurea* (see below); perennials with leaves lobed or not; not spiny. Flower-heads borne on long stalks that are swollen beneath the involucrum.

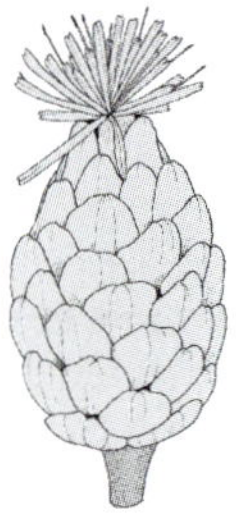

L. conifer

Cheirolophus sempervirens A knapweed-like, roughly hairy perennial to 40 cm, somewhat woody at the base with **branches leafy throughout**. Leaves lanceolate, stalkless and long-pointed. Involucre to 20 mm with hairless bracts; florets purple. Local in rocky places.

Leuzea conifer PINECONE THISTLE A small, knapweed-like, white-hairy perennial to 30 cm with stems leafy at the top. Leaves ovate to lanceolate, undivided or pinnately lobed below; divided into narrow segments above, white-felted beneath. Flower-heads solitary, to 50 mm, purple or whitish with a **prominent pinecone-like involucrum** and inconspicuous florets. Local in pine woods and scrub.

M. salmantica

Mantisalca salmantica A knapweed-like, erect, more or less hairy perennial. Leaves crowded at the base in a rosette; ovate in outline and pinnately lobed; upper leaves few, linear and toothed. Flower-heads mauve, to 20 mm, solitary or in lax branched clusters; involucre prominent and swollen in appearance, to 15 mm in diameter, with ovate, black-tipped bracts, each with a short apical spine. Cultivated ground, roadsides and dry, scrubby places.

Centaurea CORNFLOWERS AND KNAPWEEDS A large genus of annual or perennial herbs with alternate leaves divided or not. Flower-heads few, with tubular, usually purplish florets, the outermost spreading; involucre rounded to cylindrical and prominent, each bract with an apical appendage. Pappus usually feathery.

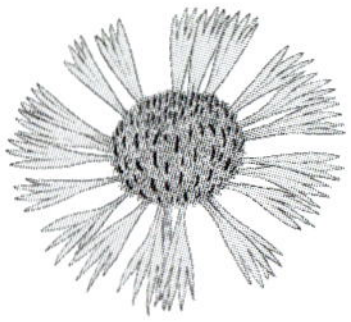

C. pullata

Centaurea pullata A short, branched or unbranched annual. Leaves roughly hairy forming a basal rosette below; lyre-shaped to slightly lobed below; pinnately lobed above. Flower-heads mauve to bluish, or white, and **large** (to 50 mm across), solitary, often surrounded by leaves; outer florets much exceeding the inner and widely spreading; flower-bracts hairless, with a conspicuous black margin and **comb-like apex**. Very common in dry, rocky, stony and sandy habitats across the region.

1

2

3

C. sphaerocephala

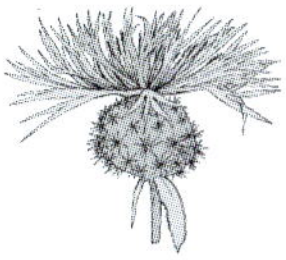

C. sphaerocephala subsp. *polycantha*

Centaurea sphaerocephala Rather similar to *C. pullata*; a medium, bushy, branched annual to 40 cm. Leaves cobweb-hairy, lyre-shaped to slightly lobed below and pinnately lobed above; stalked below but clasping the stem above. Flower-heads mauve sometimes whitish in the centre, to 35 mm across, solitary; **outer rays spreading**; flower-bracts with 5–9 yellowish **apical spines 3–5 mm long or more**. Common on coastal sands. The following two subspecies occur in the region and are sometimes treated as distinct species: **subsp. *lusitanica*** has taller, **obtusely branched stems** to 60 cm, and smaller flower-heads with involucral bracts that have yellow-orange spines; **subsp. *polycantha*** has **simple** or sparingly branched stems and an involucre to 35 mm in diameter, which is abruptly truncated at the base. **Centaurea eriophora** is similar to *C. sphaerocephala* but with flower bracts **densely clothed in hairs** (arachnoid) and with lateral appendage spines less than 5 mm long.

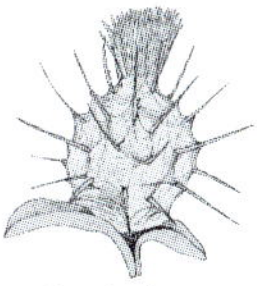

C. eriophora

Centaurea aspera Similar to *C. sphaerocephala* but with **outer ray florets scarcely spreading**, and **short apical spines on flower-bracts** (not more than 3 mm long). Dry open habitats.

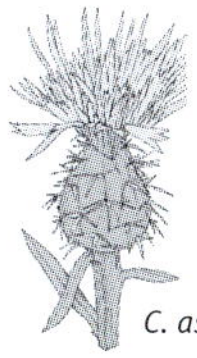

C. aspera

Centaurea calcitrapa RED STAR-THISTLE A medium, much-branched, virtually hairless perennial with grooved stems. Leaves grey when young, glandular, remotely pinnately lobed with bristle-pointed lobes, often withered below when in flower; upper leaves smaller and narrower. Flower-heads purple, to 10 mm with equal florets, surrounded by conspicuous spreading, long-spined upper leaves, forming a star. Involucre ovoid-cylindrical. Pappus absent. Waste places, roadsides, bare, sandy ground and other disturbed habitats.

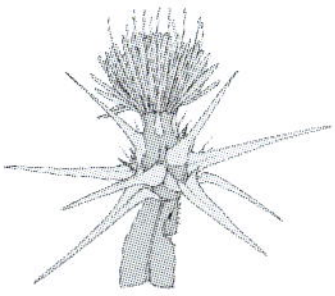

C. calcitrapa

Centaurea ornata A tall, slender and sparingly branched perennial to 80 cm tall. Leaves more or less hairless, undivided to twice pinnately lobed with narrow segments to 3 mm wide. Flower-heads dull **yellow**, often tinged orange or brown; involucre spherical, to 25 mm in diameter, **bracts with long, spreading apical spines** to 35 mm long. Dry, scrubby places. **Centaurea prolongi** is similar but with **entire**, long-stalked leaves, a smaller involucre (15–18 mm in diameter), and deep-golden to orange florets. Rocky scrub.

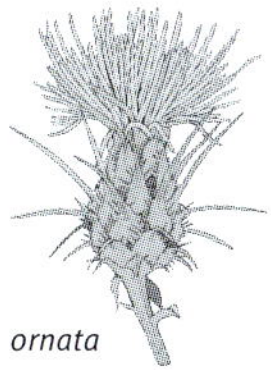

C. ornata

Carthamus Thistle-like, very spiny, often glandular annuals or perennials. Leaves pinnately lobed with spiny margins. Flower-heads solitary, surrounded by spiny, leaf-like bracts. Pappus present.

Carthamus lanatus A robust, very spiny, thistle-like annual. Stems straw-coloured, unbranched below, branched above; covered in white-woolly hairs when young. Leaves lanceolate, pinnately lobed with a spiny margin; withered below when in flower, clasping the stem above. Flower-heads to 30 mm, yellow; flower-bracts with a spine-toothed appendage. Fairly common in bare, dry and sandy places.

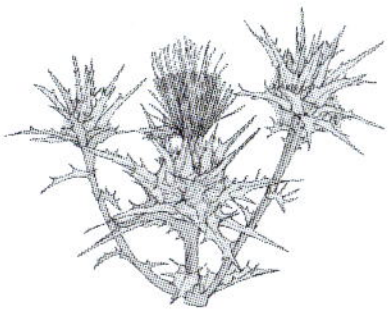

C. lanatus

1. *Centaurea pullata*
2. *Centaurea calcitrapa*
3. *Centaurea sphaerocephala*

Carduncellus caeruleus A medium, greyish, hairy perennial with normally unbranched, erect, unwinged stems. Leaves rather shiny, grey-green, lyre-shaped and toothed or untoothed to pinnately lobed with bristle-tips; upper leaves semi-clasping the stem. Flower-heads blue-purple, solitary, to 30 mm across, surrounded by leafy bracts; florets long and tubular, and **deeply 5-lobed**. Fallow land. ***Carduncellus monspelliensium*** is similar but often shorter, with fewer (up to 6) stem leaves that have 6–9 (not more than 10) lobes and flower-heads that are often stalkless. Dry, stony habitats.

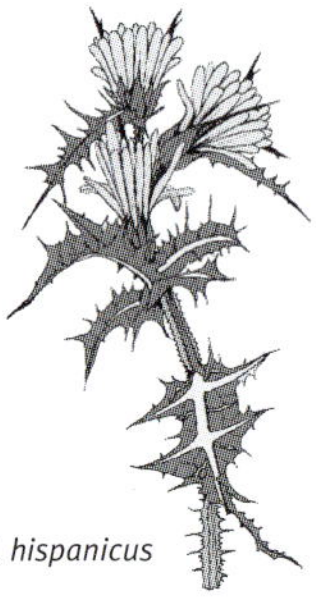
S. hispanicus

Scolymus hispanicus **Spanish oyster plant** A robust, spiny, medium biennial or perennial, superficially similar to *Carduncellus caeruleus*, but with interrupted **spiny-winged stems**. Lower leaves yellowish-green, oblong, pinnately lobed with sparse spines; upper leaves smaller and spinier. Flower-heads golden-yellow, to 30 mm, borne in a **long, narrow, spike-like panicle**; **florets rayed** (not tubular); flower-bracts with membranous margins, narrowed into a sharp point. Fairly frequent on fallow ground, roadsides and sandy waste ground.

Cichorium Annual or perennial herbs with a white latex when cut. Leaves toothed or lobed. Flower-heads numerous; florets all rayed and toothed at the tip. Pappus comprising a series of short scales.

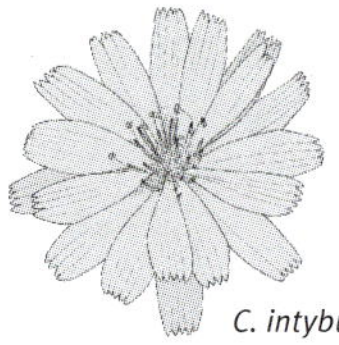
C. intybus

Cichorium intybus **chicory** A hairless to stiffly hairy, erect, branched perennial. Basal leaves pinnately lobed and short-stalked below, lanceolate and clasping the stem above. Flower-heads bright **sky-blue**, to 40 mm, borne in narrow, leafy, branched spikes. Fairly common on fallow ground and roadsides.

Tolpis Annual to perennial herbs with yellow, daisy-like flower-heads. Pappus with short hairs, long hairs, or both.

T. barbata

Tolpis barbata **tolpis** A variably sized, somewhat hairy annual, with slender, spreading, branched stems. Leaves linear-lanceolate, toothed, lobed or untoothed. Flower-heads lemon yellow with a contrasting dark purple-brown centre, to 30 mm across, borne on thickened stalks; florets all rayed. Pappus with rigid hairs. Common in disturbed sandy places, mostly in coastal areas in the region. **Var. *umbellata*** is variously regarded as distinct, with flower-heads with **all pale yellow florets**.

Hyoseris scabra A low, dandelion-like annual with a rosette of basal leaves. Leaves pinnately divided with **backward-pointing triangular lobes**. Flower-heads borne on ascending (not erect) stalks to 70 mm tall, swollen at or above the middle. Flower-heads yellow and dandelion-like, to 35 mm across; florets all rayed. Common on grassy fallow land.

1. *Scolymus hispanicus*
2. *Hyoseris scabra*
3. *Hedypnois arenaria*
4. *Hypochaeris radicata*

H. cretica

Hedypnois cretica A variable, more or less hairy annual to 45 cm tall, with mostly basal leaves and **many, branched stems**. Leaves narrowly elliptical, entire to deeply lobed, with winged stalks below, stalkless above. Flower-heads yellow and dandelion-like, borne on thickened stalks; involucre with narrow linear-lanceolate **bracts strongly incurved when in fruit**. Fairly common in a range of dry and disturbed habitats across the region. ***Hedypnois arenaria*** is similar but with flower-stalks scarcely thickened and involucral **bracts not, or only slightly incurved in fruit**. Coastal sands; local.

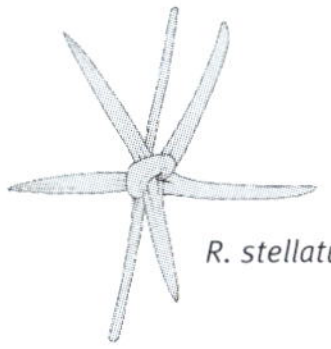

R. stellatus

Rhagadiolus stellatus STAR HAWKBIT A rather weedy, coarsely hairy annual with branched, spreading stems to 50 cm. Leaves oblong, toothed to lobed and indistinctly stalked. Flower-heads yellow, small to 10 mm across, long-stalked in lax panicles. **Fruit star-shaped** with long, slender lobes (achenes). Fallow land.

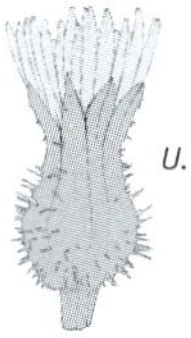

U. picroides

Urospermum picroides An annual with bristly stems. Leaves bristly, rather large to 20 cm long and sow thistle-like, toothed to lobed, oblong below and linear-lanceolate above. Flower-heads yellow and dandelion-like, large (to 40 mm across), borne on stalks that are thickened below the capitulum. Pappus white and fluffy. Local on waste ground.

Hypochaeris Annual to perennial herbs with rosettes of leaves and solitary (or few) branched stems with yellow, dandelion-like flower-heads, with rayed florets and membranous scales between.

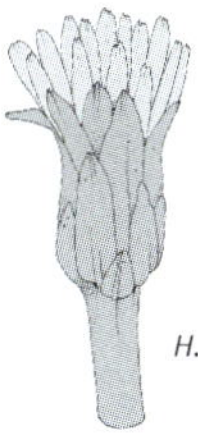

H. radicata

Hypochaeris radicata COMMON CAT'S-EAR A low perennial with almost **hairless, branched stems**. Leaves in a basal rosette, bristly hairy, wavy-toothed, oblong, to 25 cm long. **Flower-stalks thickened below the capitula** and with sparse, dark-tipped leaves. Flower-heads yellow and dandelion-like, to 40 mm across, with a bell-shaped involucre narrowed abruptly into the stem, with numerous, dark-tipped, hairless bracts. Common in grassy habitats. ***Hypochaeris glabra*** is similar but with virtually hairless, glossy leaves, and **small** (to 15 mm across), partially closed flower-heads with ray florets scarcely exceeding the bracts. Local in maritime, grassy habitats.

Leontodon taraxacoides LESSER HAWKBIT A bristly white-hairy perennial herb with leaves in a basal rosette, without tuberous roots. Leaves wavy-toothed to pinnately lobed, sparsely bristly with **forked bristles**. Flowerheads solitary on **unbranched, leafless stalks** drooping in bud, to 25 mm across; bracts hairless except along the midribs. Achenes with a pappus of scales. Common in grassy habitats. ***Leontodon tuberosus*** is similar but with **tuberous roots** and **achenes with a pappus of short hairs**. Habitats similar to those of *L. taraxacoides*. ***Leontodon salzmannii*** is a similar, **hairless** species that is poorly known, and rare in Europe, but that might occur in the region in sandy places.

1. *Leontodon taraxacoides*
2. *Picris algarbiensis*
3. *Picris echioides*
4. *Tragopogon hybridus*
5. *Reichardia gaditana*
6. *Reichardia gaditana*
7. *Aetheorhiza bulbosa* leaves
8. *Aetheorhiza bulbosa*

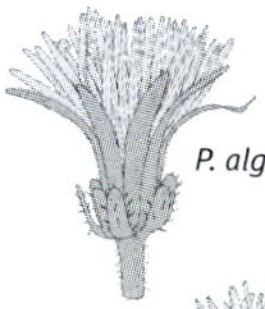

P. algarbiensis

P. comosa

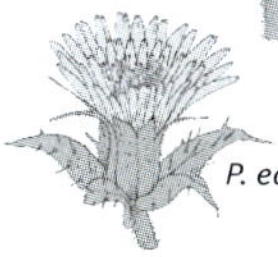

P. echioides

Picris algarbiensis A very rough-hairy biennial to perennial with solitary, branched stems to 90 cm tall. Leaves with scattered **prominent spines with bulbous, pronounced bases.** Flower-heads with yellow, rayed florets; involucre long, to 18 mm, with **inner involucral bracts 3 times the length of the outer**, linear-oblong. **Achenes with a long beak, to twice as long as the body**; pappus white. *Matos*, roadsides and woods on higher ground; an endemic. ***Picris comosa*** is similar but normally an annual, with **longer involucre** (to 25 mm) **with inner bracts more than twice as long as the outer**, and **achenes with a much shorter beak (to half as long as the body)**; dry scrub. ***Picris echioides*** is similar to both the above species but with shorter involucre (12–20 mm long), with **inner involucral bracts less than twice the length of the outer**. Rather casual and sporadic in the region in dry, scrubby waste places.

Tragopogon **GOAT'S BEARDS** Annual to perennial herbs with a white latex when cut. Stems usually solitary. Leaves linear, often rush-like. Flower-heads with all rayed florets. Fruit a large 'dandelion clock'.

T. crocifolius

Tragopogon hybridus A slender, hairless annual to 60 cm with branched stems. Leaves long-linear and rush-like (not typical of the family). Flower-heads borne on inflated stalks; lilac pink, with a few, purplish central florets much exceeded by the surrounding rays, which are in turn exceeded by the narrow, long-pointed flower bracts. Common on stony fallow land. ***Tragopogon crocifolius*** is similar but with darker purple or brownish flower-heads with **yellow central florets**, and the ray florets less markedly exceeded by the flower bracts. Local in stony and grassy waste places.

R. picroides

Reichardia gaditana Superficially like a sow thistle; a short but robust, hairless annual to 35 cm tall. Leaves smooth and hairless with spine-toothed margins, narrowed at the base into a winged stalk. Involucre prominent and bell-shaped, to 30 mm, with large bracts with **wide**, somewhat wavy, pale brown, membranous margins. Flower-heads yellow and dandelion-like; the rays yellow and **purple at the base. All achenes wrinkled**. Common on coastal sands and in maritime grassy habitats. ***Reichardia picroides*** is a perennial that is similar to *R. gaditana* but with the flower bracts having only a **narrow** membranous margin (less than 0.5 mm in width), the **florets all yellow (not purplish at the base)** though sometimes with a dark stripe on the outer surface, and the outer achenes wrinkled, but the **inner smooth**, and appearing sterile. Cultivated and waste land. ***Reichardia intermedia*** is an annual with flower bracts that have wide margins (to 1.25 mm) as in *R. gaditana*, but the **innermost achenes smooth**, and appearing sterile like *R. picroides*. Cultivated and on waste land.

Aetheorhiza bulbosa A small, purplish perennial with stolons, rhizomes and few, slender stems. **Leaves held prostrate on the surface of the sand,** all basal, hairless or hairy, elliptical and entire or lobed. Flower-heads borne terminally; yellow and dandelion-like; involucral bracts with blackish hairs. Common on coastal sands, rare elsewhere.

Sonchus **SOW THISTLES** Annual or perennial herbs with stout, hollow stems exuding a white, sticky latex when cut. Leaves pinnately lobed or wavy with spiny margins; the stem leaves clasping at the base. Flower-heads yellow, the florets all rayed.

S. asper

Sonchus asper **PRICKLY SOW-THISTLE** A variably tall, erect, greyish or reddish, hairless (glandular above) annual to 1 m, with simple or branched stems. Lower leaves spoon-shaped, sometimes pinnately lobed, with triangular toothed lobes, the **lowermost two rounded at the base**, and **glossy** green on the upper surface; clasping the stem above. Flower-heads golden yellow, to 25 mm in diameter. Very common in a wide range of habitats from towns to coastal dunes across the region. ***Sonchus tenerrimus*** is similar to *S. asper* but with deeply pinnately divided leaves with all leaf lobes **strongly constricted or linear at the base**, and the terminal lobe equalling the laterals (not distinctly larger). Flower-stalks often white-hairy. Local on disturbed habitats inland.

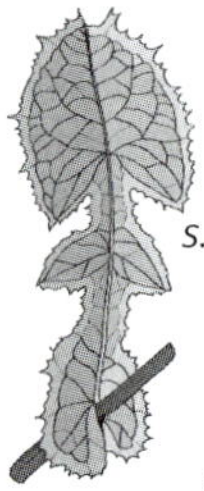

S. oleraceus

Sonchus oleraceus **SMOOTH SOW-THISTLE** Rather similar to *S. asper*; an erect, greyish or reddish, hairless (glandular above) annual to 1 m. Lower leaves spoon-shaped and usually **pinnately lobed** with triangular toothed lobes, the **end-lobe distinctly larger than the next pair down** and the **lowermost two pointed** (not rounded) at the base, **matt** (not glossy) green on the upper surface; leaves clasping the stem above. Flower-heads pale yellow, to 25 mm in diameter. Very common in a range of habitats similar to those of *Sonchus asper*.

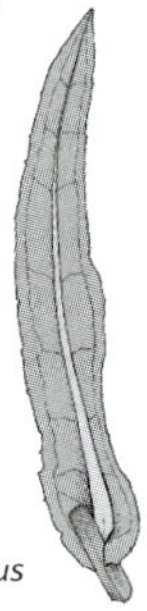

S. maritimus

Sonchus maritimus Similar to other *Sonchus* spp. but a **short, rhizomatous perennial** to 60 cm tall or less, with simple or sparingly branched stems. Leaves mostly at the base, linear and wavy-toothed or rarely lobed, the basal lobes rounded. Flower-stalks often **white-downy**. Local on damp coastal sands.

Lactuca **LETTUCES** Annual, biennial and perennial herbs with a white latex when cut. Flower-heads with cylindrical involucre and all florets rayed.

L. serriola

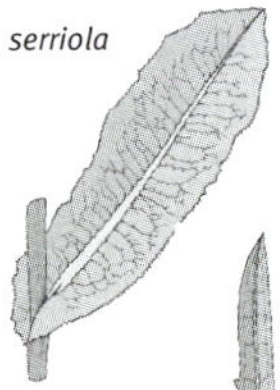

Lactuca serriola A tall, greyish, stiffly erect annual or biennial to 2 m, unbranched below, and branched above. **Leaves held stiffly erect**, oblong-lanceolate, often pinnately divided with distant lobes below, more or less hairless but spiny along the midrib beneath and along the margins; all leaves waxy and greyish. Flower-heads pale yellow and small, to 13 mm across, borne in a long, lax inflorescence. Local on waste ground in the region. ***Lactuca saligna*** is similar but with **whitish stems to 1 m or less** and greyish **leaves not prickly**. Bare, grassy habitats.

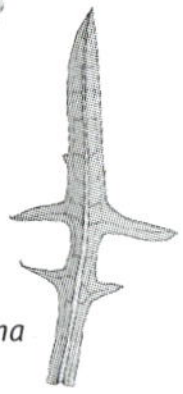

L. saligna

C. juncea

Chondrilla juncea A medium to tall greyish biennial, hairy, especially below. Stems normally solitary, to 1 m tall, stiff and broom-like with few leaves. Leaves oblong, lobed, withering below and linear and entire above. Flower-heads unstalked, in small clusters; florets all rayed. Sandy or scrubby open habitats.

1. *Sonchus asper*
2. *Sonchus oleraceus*
3. *Taraxacum officinale*

Taraxacum **DANDELIONS** A taxonomically complex group of perennial herbs with tap roots and a basal rosette of leaves; flower-heads solitary on leafless, hollow stems that exude a milky latex when cut.

Taraxacum officinale **COMMON DANDELION** A variable perennial with leaves in a basal rosette. Leaves pinnately lobed, with long, winged stalks at the base. Flower-heads few to many, solitary on leafless, hollow stalks to 40 cm tall; bright yellow and typically dandelion-like, to 60 mm across; outer flower bracts backwardly curved. Fruits borne in a 'dandelion clock'. Common in grassy habitats across the region.

Crepis **HAWK'S-BEARDS** Annual or perennial herbs with spirally arranged leaves with lobes pointing backwards, and erect, branched stems. Flower-heads yellow with florets all rayed; flower bracts in 2 rows, the outer row often shorter and spreading. Pappus consisting of white, brittle hairs.

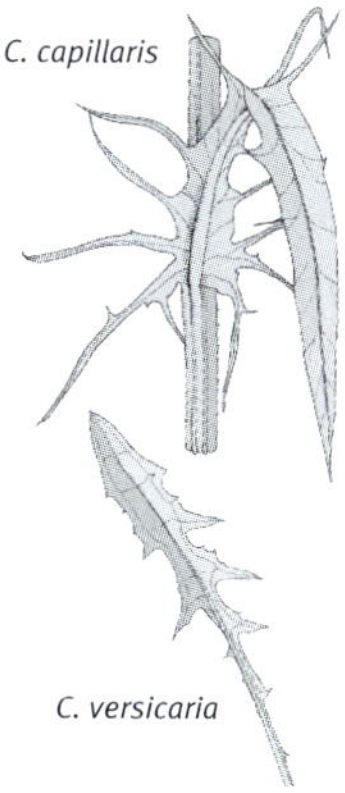

C. capillaris

C. versicaria

Crepis capillaris **SMOOTH HAWK'S-BEARD** An erect, hairless annual to 1 m, branched from the base or above. Leaves rather hairy, glossy, irregularly, narrowly pinnately lobed, with a large triangular end-lobe and narrower lateral lobes; upper leaves smaller and stalkless with **clasping, arrow-shaped bases**. Flower-heads in loose inflorescences, held erect in bud; yellow and dandelion-like, to 15 mm across; involucral bracts downy, often with blackish hairs. Widespread in disturbed waste places. ***Crepis versicaria*** is similar but **downier**, the leaves with broader lobes and downy all over. Flower-heads with **orange-yellow florets**, the outer striped reddish externally; outer involucral bracts spreading. **Achenes with long, slender beaks**. Habitats similar to those of *C. capillaris*.

A. integrifolia

Andryala integrifolia A variable, short to tall, white-hairy annual to perennial with sparingly branched, erect, leafy stems. Leaves ovate to lanceolate, unlobed to lobed, semi-clasping the stem above, and densely covered in yellowish glandular hairs. Flower-heads pale lemon yellow, to 20 mm across, borne in loose rather flat-topped clusters; flower bracts linear-lanceolate and hairy. Locally common in dry, sandy and grassy habitats. ***Andryala laxiflora*** has larger flower-heads with an involucre to 15 mm wide (not 5–10 mm). ***Andryala ragusina*** is white-felted but with an **involucre without glandular hairs**.

Glossary

Achene, a simple, singled-seeded, dry indehiscent fruit; there are often many achenes on a fruiting head, as in *Ranunculus*.

Actinomorphic, (of a flower) with a radially symmetrical (regular) shape that has multiple axes of symmetry, e.g. a lily flower.

Alien, not native, introduced and now naturalised.

Alternate, (of leaves) arising at different heights along the stem (not opposite or whorled).

Annual, completing the life cycle in one year.

Anther, the fertile, pollen-producing part of the stamen (syn. microsporangium).

Apex, the tip or uppermost part of a structure.

Ascending, arising upwards at an angle.

Awn, long, stiff bristle.

Axil, the point at which the leaf or leaf stalk joins the stem; adj. axillary.

Beak, an elongated projection, usually on a fruit; adj. **beaked**.

Biennial, completing its life cycle in two years.

Bipinnate, divided twice (doubly pinnate): leaves divided into segments, which themselves are divided into segments.

Blade, the flattened part of the leaf.

Bract, small, leaf-like structure, subtending (beneath) a flower or inflorescence.

Bracteole, small bract.

Bulb, underground storage organ composed of fleshy leaves.

Bulbil, a small reproductive bulb borne in the leaf axil or within flowers.

Calyx, all the sepals of the flower.

Capitulum, a flower-head; dense inflorescence composed of small stalkless flowers crowded together on a compound receptacle; typical of the daisy family (Asteraceae).

Carpel, the female organ of the flower, comprising stigma, style and ovary.

Casual, introduced to an area and ephemeral in its appearance.

Compound, made up of more than one similar part or segment, typically a leaf.

Cordate, heart-shaped.

Corm, underground storage organ formed from a swollen stem base.

Corolla, all the petals of a flower (which may form a tube).

Cosexual, flowers with both male and female reproductive organs, stamens and carpels, respectively.

Cyathium, specialised inflorescence of *Euphorbia* (Euphorbiaceae) with a cup-like structure containing a single carpellate (female) flower and several staminate (male) flowers.

Deciduous, falling off, for example leaves in autumn.

Desiccation, drying out.

Dioecious, with male and female flowers on separate plants (individual plants of one sex).

Disc floret, small actinomorphic flower, forming part or all of a capitulum (in Asteraceae).

Endemic, restricted to a particular country or area.

Entire, whole; without lobes or divisions.

Epicalyx, additional whorl of sepal-like bracts beneath the true sepals.

Falls, outer perianth segments (petals) of *Iris* flowers (Iridaceae).

Family, a monophyletic group of related genera, the taxonomic group between the lower rank of genus and the higher rank of order.

Floret, a small individual flower making up part of a dense inflorescence, e.g. part of the capitulum of Asteraceae species.

Fruit, the ripened ovary or ovaries of a flower containing seeds.

Genus (pl. **genera**)**,** a monophyletic group of related species, the taxonomic group between the lower rank of species and the higher rank of family; the generic name is the first part of the scientific binomial.

Geophyte, a plant that survives the dry summer as a dormant underground bulb, corm or tuber.

Glabrous, not hairy.

Gland, organ of secretion, often in sticky plants; adj. **glandular**.

Glaucous, covered in a bluish, whitish or greyish waxy bloom.

Globose, spherical.

Halophyte, salt-tolerant plant.

Head, group of flowers crowded together at the end of a stalk.

Hemiparasite, parasitic plant that gains some of its nutrition from another plant (host) but which also has chlorophyll and a root system; some hemiparasites can survive independently of a host (facultative hemiparasitism).

Herb, a plant without woody parts; a soft and leafy annual, biennial or perennial whose aerial parts naturally die to ground level at the end of the growing season; adj. **herbaceous**.

Hermaphrodite, organism or structure bearing both male and female reproductive organs. Hermaphrodite flowering plants can be monoecious, i.e. with unisexual male and female flowers on the same plant, or co-sexual (syn. bisexual) with male stamens and female carpels within the same flower.

Holoparasite, parasitic plant which gains all of its nutrition from another plant (the host) and which lacks chlorophyll and a true root system.

Hybrid, the offspring of a cross between two different species, races, or varieties.

Inferior ovary, an ovary that is situated beneath the point of insertion of other floral organs; syn. **epigynous ovary**.

Inflorescence, a group of flowers on a floral stem (axis).

Internode, a part of the stem between two nodes.

Involucre, collection of involucral bracts.

Involucral bract, bracts surrounding a head of flowers (e.g. in Asteraceae).

Keel, boat-shaped structure formed by two lower petals in the pea family (Leguminosae).

Labellum, lower-most petal of an orchid flower, often highly specialised, e.g. in bee orchids (*Ophrys* spp.).

Lanceolate, spear-shaped.

Lemma, (of grasses) the lower of the pair of bracts (lemma and palea) that subtends the floret.

Ligule, (of grasses) a small membranous projection or ring of hairs at the junction of the leaf sheath and stem.

Linear, narrow and parallel, often grass-like.

Maquis, thicket of shrubby sclerophyllous vegetation in the Mediterranean.

Matos, Portuguese for maquis.

Membranous, paper-like.

Monoecious, with separate male and female reproductive structures on the same individual plant; contrast dioecious.

Mycoheterotrophy, process by which a non-photosynthetic plant obtains nutrition from a fungal symbiont (or sometimes from another plant via a shared fungal symbiont) living in its root system.

Native, naturally occurring in the area.

Naturalised, not native but well established.

Oblong, elongated but wide in shape (usually describing a leaf).

Ovary, part of the carpel or pistil containing the ovules, and later, the seeds.

Ovate, egg-shaped.

Ovule, structure containing the egg, becomes the seed after fertilisation.

Palmate, lobes or segments radiating from a common axis.

Panicle, branched inflorescence.

Pappus, a structure consisting of hairs, bristles or scales on the fruit of Asteraceae.

Parasite, plant obtaining nutrients from another plant (may be hemi- or holo-parasitic).

Pedicel, the stalk of an individual flower in an inflorescence, or the stalk of a grass spikelet.

Perennial, living for more than two years, generally flowering every year at maturity.

Perfoliate, (of a leaf or bract) with the base united (around the stem).

Perianth, all the sepals and petals of a flower.

Pinnate, (of a compound leaf) composed of leaflets arranged on opposite sides of a common axis (rachis); adverb. **pinnately**.

Pinnatifid, (of a leaf) pinnately divided with the lobes cut nearly (but not quite) to the mid-vein.

Pollinium (pl. **pollinia**), a mass of adhering pollen grains (of e.g. an orchid) that is shed and transported as a unit by a pollinator.

Procumbent, trailing on the ground.

Pseudocopulation, the process by which a male insect (usually a bee or wasp) attempts to mate with the flower of a bee orchid (*Ophrys* spp.) and in so doing, brings about cross-fertilisation.

Raceme, a simple unbranched inflorescence with stalked flowers borne on a single axis, the youngest flowers at the top; adj. **racemose**.

Rachis, stalk of a compound leaf or the central axis bearing the flowers.

Ray, a radiating branch of an umbel.

Ray floret, a small zygomorphic flower often resembling a single petal in the inflorescence (head) of Asteraceae. Contrast with disc floret.

Receptacle, the portion of the axis of a flower stalk on which the flower is borne.

Rhizome, horizontal underground stem.

Rosette, (of leaves) radiating from a central point on the ground.

Sclerophyll, a hard leathery leaf containing a high proportion of thickened cells (sclereids); adj. **sclerophyllous**.

Sepal, leaf-like segments of the perianth (occasionally like a petal and then called a tepal).

Simple, structure (e.g. a leaf) that is not divided into segments or lobes.

Spadix, a spike-like organ bearing tiny male and female flowers at its base and surrounded by a spathe; characteristic of Araceae.

Spathe, large, leafy bract, sometimes brightly coloured, enclosing a spadix; characteristic of Araceae.

Spike, an unbranched inflorescence of stalkless flowers, the youngest at the top.

Spikelet, the inflorescence of grasses (Poaceae) and sedges (Cyperaceae).

Spur, hollow, tubular projection originating from the sepals or petals, often containing nectar.

Stamen, male reproductive organ of the flower consisting of a filament and anther (microsporangium).

Standard, upper petals of flowers of species in the pea family (Leguminosae) and iris family (Iridaceae).

Stellate, star-shaped.

Stigma, the part of the carpel or pistil that receives pollen and upon which the pollen germinates.

Stipule, small, leaf-like organ at the base of some leaf petioles (stalks), often in pairs, may be simple or lobed.

Subspecies, a taxonomic subdivision of a species, usually a distinct geographical race.

Superior ovary, ovary positioned above the attachment points of other floral organs; syn. hypogynous ovary.

Taxon (pl. **taxa**), a taxonomic unit of any rank, for example species, genus, subspecies or variety.

Teeth, small, triangular segments; adj. **toothed**.

Trifoliate, with three leaflets.

Umbel, a flat-topped or convex 'umbrella-shaped' inflorescence consisting of a cluster of flowers with spreading stalks (pedicels) that arise from the apex of the peduncle; typical of Apiaceae.

Unisexual, flowers with either male or female reproductive organs only. Contrast cosexual.

Variety, a taxon that shows distinct characteristics or is geographically distinct (less distinct than a subspecies) within a defined species.

Viscid, sticky.

Xerophyte, drought-tolerant plant.

Zygomorphic, (of a flower) with a bilaterally symmetrical, irregular shape.

Index of English names

T

Index of scientific names

Page numbers in **bold** indicate a photograph.

C

D